AF361795

Rainfall Infiltration Modeling

Rainfall Infiltration Modeling

Special Issue Editors

Renato Morbidelli
Carla Saltalippi
Alessia Flammini

MDPI • Basel • Beijing • Wuhan • Barcelona • Belgrade

Special Issue Editors

Renato Morbidelli
Universita degli Studi di
Perugia, Department of
Civil & Environmental
Engineering
Italy

Carla Saltalippi
Universita degli Studi di
Perugia, Department of
Civil & Environmental
Engineering
Italy

Alessia Flammini
Universita degli Studi di
Perugia, Department of
Civil & Environmental
Engineering
Italy

Editorial Office
MDPI
St. Alban-Anlage 66
4052 Basel, Switzerland

This is a reprint of articles from the Special Issue published online in the open access journal *Water* (ISSN 2073-4441) from 2018 to 2020 (available at: https://www.mdpi.com/journal/water/special_issues/Rainfall_Infiltration_Modeling).

For citation purposes, cite each article independently as indicated on the article page online and as indicated below:

LastName, A.A.; LastName, B.B.; LastName, C.C. Article Title. *Journal Name* **Year**, *Article Number*, Page Range.

ISBN 978-3-03936-022-2 (Hbk)
ISBN 978-3-03936-023-9 (PDF)

Contents

About the Special Issue Editors

Renato Morbidelli has served as Associate Professor of Hydrology and Hydraulic Works at the University of Perugia in the Department of Civil and Environmental Engineering since 2002. He reached enablement for the national position of Full Professor of Hydrology and Hydraulic Works in 2016. He completed his PhD in 1998 at the University of Padua. From 1997 to 2002, he served at the University of Perugia as an Assistant Professor of Hydrology and Hydraulic Works. He has authored of numerous books and many papers published in prestigious international journals. The research activities of Morbidelli have been mainly focused on rainfall infiltration, the effects of spatial variability on soil hydraulic properties, and the use of simplified models for computing surface runoff and extreme rainfall analysis.

Carla Saltalippi completed her PhD at the University of Pavia in 1999. She is Associate Professor at the Department of Civil and Environmental Engineering of the University of Perugia, Italy, where she teaches hydrology-related disciplines. She is the author of many papers in prestigious international journals and her main research interests focus on rainfall–runoff modeling, infiltration modeling, real-time flood forecasting, ground water, soil moisture, and extreme rainfall analysis.

Alessia Flammini has served as Researcher at the University of Perugia, in the Department of Civil and Environmental Engineering since 2006, and as Assistant Professor of Hydrology and Transport Processes in Fluids and Soil since 2009. In 2006, she completed her PhD at the University of Perugia in Civil Engineering and Curriculum in Engineering of Water. She is author of more than 30 papers published in the main international scientific journals with subject categories of Water Resources and Environmental Sciences (Journal of Hydrology, Hydrological Processes, Water Resources Research, Advances in Water Resources), 3 book chapters, and around 20 conference proceedings. The main topics of her research are (i) rainfall infiltration modeling at local and areal scales; (ii) representation of the spatial variability of soil hydraulic properties; (iii) rainfall–runoff modeling; and (iv) soil moisture trends in space and time.

Preface to "Rainfall Infiltration Modeling"

Rainfall infiltration plays a fundamental role within the hydrologic cycle. The spatiotemporal evolution of the infiltration rate under natural conditions cannot currently be deduced by direct measurements at any scale of interest in applied hydrology and, therefore, the use of infiltration modeling is of crucial importance in allowing it to be described through measurable quantities. In spite of continuous developments in infiltration modeling, the estimate of infiltration at different spatial scales, i.e., from the local to watershed scales, is a complex problem because of the natural spatial variability of both soil hydraulic characteristics and rainfall. For many years, research activity has been limited to the development of local or point infiltration models for vertically homogeneous soils. However, in addition to deepening this modeling, other interesting open problems should be addressed, including the modeling of point infiltration into vertically non-uniform soils, infiltration over horizontal heterogeneous areas, and infiltration into soil surface with significant slopes. The main objective of this Special Issue concerning the rainfall infiltration modeling is to put together updated and original contributions on this topic as a basic element to emphasize the critical points that require substantial research development. This Special Issue consists of 10 papers, including a review paper focused on rainfall infiltration modeling, which retraces some important milestones that led to the definition of basic mathematical models both at the local and field scale. In the same paper, some open problems involving the vertical and horizontal inhomogeneity of the soils are explored, while rainfall infiltration modeling over surfaces with significant slopes is considered. In the remaining papers, a wide range of topics and research questions is discussed. More specifically, a new conceptual model for slope infiltration and the effect of wastewater in the infiltration process has been developed, along with different recharge models and original sensitivity analyses in the field of groundwater.

Renato Morbidelli, Carla Saltalippi, Alessia Flammini
Special Issue Editors

Review

Rainfall Infiltration Modeling: A Review

Renato Morbidelli [1,*], Corrado Corradini [1], Carla Saltalippi [1], Alessia Flammini [1], Jacopo Dari [1] and Rao S. Govindaraju [2]

[1]   Department of Civil and Environmental Engineering, University of Perugia, via G. Duranti 93, 06125 Perugia, Italy; corrado.corradini@unipg.it (C.C.); carla.saltalippi@unipg.it (C.S.); alessia.flammini@unipg.it (A.F.); jacopo.dari@unifi.it (J.D.)
[2]   Lyles School of Civil Engineering, Purdue University, West Lafayette, IN 47907, USA; govind@purdue.edu
*   Correspondence: renato.morbidelli@unipg.it; Tel.: +39-075-5853620

Received: 7 December 2018; Accepted: 13 December 2018; Published: 18 December 2018

Abstract: Infiltration of water into soil is a key process in various fields, including hydrology, hydraulic works, agriculture, and transport of pollutants. Depending upon rainfall and soil characteristics as well as from initial and very complex boundary conditions, an exhaustive understanding of infiltration and its mathematical representation can be challenging. During the last decades, significant research effort has been expended to enhance the seminal contributions of Green, Ampt, Horton, Philip, Brutsaert, Parlange and many other scientists. This review paper retraces some important milestones that led to the definition of basic mathematical models, both at the local and field scales. Some open problems, especially those involving the vertical and horizontal inhomogeneity of the soils, are explored. Finally, rainfall infiltration modeling over surfaces with significant slopes is also discussed.

Keywords: hydrology; infiltration process; local infiltration models; areal-average infiltration models; layered soils

---

1. Introduction

Brutsaert [1] offers a concise definition of infiltration as "the entry of water into the soil surface and its subsequent vertical motion through the soil profile". Infiltration plays an important role in the partitioning of applied surface water into surface runoff and subsurface water—both of these components govern water supply for agriculture, transport of pollutants through the vadose zone, and recharge of aquifers [2]. The infiltration of rain and surface water is influenced by many factors, including soil depth and geomorphology, soil hydraulic properties, and rainfall or climatic properties [3]. Researchers have understood for ages that rainfall wets the soil and may produce runoff. Our understanding of the physics of the process and the dynamics of porous media hydraulics has come rather recently. Our ability to mathematically describe the response of a soil to rainfall and to understand the parameters that affect infiltration, has developed only in the last few decades.

The spatio-temporal evolution of infiltration rate under natural conditions cannot be currently deduced by direct measurements at all scales of interest in applied hydrology, and infiltration modeling with the aid of measurable quantities is of fundamental importance.

Even though the representation of the main natural processes in applied hydrology requires areal infiltration modeling for both flat and sloping surfaces, research activity has been limited to the development of local or point infiltration models for many years. A variety of local infiltration models for vertically homogeneous soils with constant initial soil water content and over horizontal surfaces has been proposed [4–28]. A milestone paper describing the process of infiltration from a ponded surface condition was published by Reference [4]. Successively, Reference [29] published his assessment of the role of infiltration in flood generation, defining "infiltration capacity" (*IC*) as a hyetograph separation rate that was generally applicable as a threshold for application to a rainfall intensity graph. A few years later, Reference [30] refined this concept by referring to it as an infiltration rate that declines exponentially during a storm, and then published a conceptual derivation of the exponential decay infiltration equation. On this basis, if the rainfall overcomes the *IC*, only a portion of it may infiltrate while the remaining quantity ponds over the surface or moves depending of the local slope. Therefore, the *IC* can be regarded as an important soil characteristic. As an example, Figure 1 shows several infiltration regimes when rainfall occurs over a soil surface. The line represents the *IC*(t) curve for a silty loam soil. The dark columns represent the rainfall rate, $r(t)$, observed in the Umbria Region (Central Italy) during a generic frontal system. The dashed columns represent the simulated behavior of infiltration rate, $f(t)$, during the rainfall event. With low values of rainfall intensities, the total amount of $r(t)$ infiltrates through the soil surface and $r(t)$ is equal to $f(t)$. With high $r(t)$ values only a part of the $r(t)$ can infiltrate, while the difference $r(t)-f(t)$ becomes runoff. As it can be observed, there exists no straightforward relationship between *IC* and runoff as the effective $f(t)$ is a function of the specific rainfall intensity.

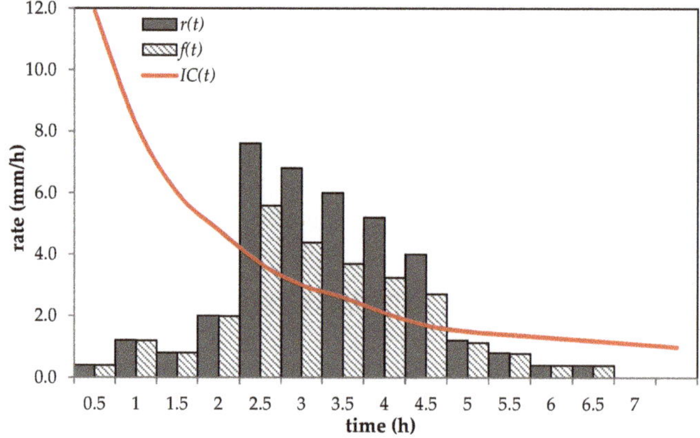

Figure 1. Infiltration capacity, *IC*(t), and infiltration rate, $f(t)$, of a silty loam soil under a natural rainfall event (observed in Central Italy) characterized by a variable intensity, $r(t)$.

Natural soils are rarely vertically homogeneous. In hydrological simulations, the estimate of effective rainfall can be reasonably schematized by a two-layered vertical profile [31,32]. Soils representable by a sealing layer over the parent soil or a vertical profile with a more permeable upper layer are found frequently in nature. Some models for infiltration into stable crusted soils were developed under significant approximations by adapting the Green–Ampt model [33–38] or the two-stage infiltration equations proposed by Mein and Larson [39] for homogeneous soils. An efficient approach which represents transient infiltration into crusted soils was proposed by Reference [40], while Reference [41] formulated a model which describes upper-layer dynamics but in the limits of a ponded upper boundary. A more general semi-analytical/conceptual model for crusted soils was later formulated by Reference [42], and was extended by Reference [43] to represent infiltration and re-infiltration after a redistribution period under any rainfall pattern and for any two-layered soil where either layer may be more or less permeable than the other. For a much more permeable upper layer, and under more restrictive rainfall patterns, a simpler semi-empirical/conceptual model was presented by Reference [44]. Under conditions of surface saturation, a simple Green–Ampt-based model was proposed [45].

In applied hydrology, upscaling of point infiltration models to the field scale is required to estimate areal-average infiltration. This is a challenging task because of the natural spatial heterogeneity of hydraulic soil properties [46–52], and particularly of the soil saturated hydraulic conductivity [53,54] that may be assumed as a random field with a lognormal univariate probability distribution.

The mathematical problem is not analytically tractable, whereas the use of accurate Monte Carlo (MC) simulation techniques imposes an enormous computational burden for routine applications. MC simulations were used, for instance, by References [55–57] to describe field-scale infiltration. MC simulations were also used by Reference [58] to corroborate a relation between areal average and variance of infiltration rate under a time-invariant rainfall rate; however, the averaging procedure was applied in space over a single realization and the behavior of the resulting errors was not specified. A variant to MC sampling for the representation of the random variability of a soil property is Latin Hypercube sampling [59] adopted by Reference [60] to develop a simple parameterized approach for areal-average infiltration.

Even though MC simulations, performed by many realizations of the random variable, are rather expensive in terms of computational effort for practical applications, they are useful as a tool to parameterize simple semi-empirical approaches or to serve as a benchmark for validating semi-analytical models. Along these lines, Reference [61] developed three versions of a semi-analytical/conceptual model for estimating the expected areal-average infiltration into vertically homogeneous soils under spatially uniform rainfall, but with random horizontal values of the saturated hydraulic conductivity (see also [62]). A shortcoming of this model is that the process of infiltration of overland flow running over pervious downstream areas (run-on process) is neglected. On the other hand, the importance of run-on was shown in a few investigations concerning the effects of horizontal variability of saturated hydraulic conductivity on Hortonian overland flow [57,63–67], but this process has generally been disregarded in hydrological models.

In addition to the heterogeneity of saturated hydraulic conductivity, rainfall is also characterized by spatial variability [68,69]. Some studies combining the random variability of these two quantities were performed by References [70–72]. The latter paper presented a semi-analytical model developed under less restrictive conditions, even though run-on was not incorporated, and was based upon the use of cumulative infiltration as the independent variable that was linked with an expected time. Subsequently, Reference [73] formulated a more complete mathematical model for the expected areal-average infiltration, which considers both the saturated hydraulic conductivity and rainfall rate as random variables, and then combines the aforementioned semi-analytical approach with a semi-empirical/conceptual component to represent the run-on process.

A model of the expected areal-average infiltration into a much more permeable upper layer of a two-layered soil was also proposed by Reference [74] considering only a spatial horizontal random field of the saturated hydraulic conductivity. It involves the solution of a set of algebraic equations obtained by upscaling simple local infiltration equations to the field scale. The areal-average infiltration for a vertically non-uniform soil characterized by a saturated hydraulic conductivity decreasing with depth according to a power law was derived by Reference [75] and upscaled to the field scale by the same semi-analytical technique used in previous papers [61,72].

Most of the above-mentioned models assumed zero or small surface slope that does not affect the infiltration process. However, in most real situations, infiltration occurs over surfaces characterized by different gradients [76,77] and the role of surface slope on infiltration is not clear. In fact, the results obtained by some theoretical and experimental investigations [78–97] lead to contrasting conclusions, suggesting that an improved understanding and modeling of infiltration on sloping surfaces is required.

Finally, when macropore flow plays a significant role in determining infiltration, amendments of the Darcy–Richards approach may be used, especially at the local scale [98,99].

The main intent of this review paper is to critically assess the complexities of the infiltration process and to provide guidance for developments related to open problems. This paper will provide classical approaches developed for rainfall events with continuous saturation at the soil surface, and a more general formulation suitable for any type of rainfall pattern for applications at the local (point) scale in homogeneous soils. Simple models for point infiltration into a two-layered soil with a more permeable upper layer and a more complex model for any two-layered soil type are presented.

Semi-empirical and semi-analytical field-scale infiltration models are analyzed. Finally, problems linked to rainfall infiltration into surfaces with significant slopes are also discussed.

2. Basic Physical Models for Infiltration

By considering a horizontally homogeneous soil, water movement in the vertical direction is governed by one-dimensional soil water flow and continuity equations. The flow rate, q, per unit cross-sectional area is described by Darcy's law, actually proposed by Reference [100], as:

$$q = -K\left(\frac{\partial \psi}{\partial z} - 1\right),\tag{1}$$

where K is the hydraulic conductivity, ψ is the soil water matric capillary head, and z is the vertical soil depth assumed positive downward. The infiltration rate, q_0, is given by Equation (1) applied at the soil surface.

In the absence of changes in the water density and soil porosity as well as of sinks and sources, the continuity equation is:

$$\frac{\partial \theta}{\partial t} = \frac{\partial q}{\partial z},\tag{2}$$

where θ is volumetric water content and t is time. The substitution of Equation (1) into Equation (2) leads to the well-known Richards equation:

$$C_1 \frac{\partial \psi}{\partial t} = \frac{\partial}{\partial z}\left[K\frac{\partial \psi}{\partial z}\right] - C_2\frac{\partial \psi}{\partial z},\tag{3}$$

where $C_1 = d\theta/d\psi$ and $C_2 = dK/d\psi$ with a typical assumption that θ and K are unique functions of ψ thus neglecting hysteresis in these functions. The initial condition at time $t = 0$ for $z \geq 0$ is $\psi = \psi_i$, and the upper boundary conditions at the soil surface, where $z = 0$, are:

$$q_0 = r, 0 < t \leq t_p; \theta_0 = \theta_s, t_p < t \leq t_r; q_0 = 0, t_r < t,\tag{4}$$

where r is the rainfall rate, t_p is the time to ponding, and t_r is the duration of rainfall. Hereafter the subscripts i and s denote initial and saturation quantities, respectively, while 0 stands for quantities at the soil surface. The lower boundary conditions at a depth z_b which is not reached by the wetting front is $\psi(z_b) = \psi_i$ for $t > 0$. The soil water hydraulic properties can be represented by the following parameterized forms [22]:

$$\psi = \psi_b \left[\left(\frac{\theta - \theta_r}{\theta_s^* - \theta_r}\right)^{-c/\lambda} - 1\right]^{1/c} + d,\tag{5a}$$

$$K = K_s^* \left[1 + \left(\frac{\psi - d}{\psi_b}\right)^c\right]^{-(b\lambda + a)/c},\tag{5b}$$

where θ_s^* and K_s^* are used as scaling quantities, ψ_b is the air entry head, θ_r is the residual volumetric water content, c, λ, and d are empirical coefficients, and $b = 3$ and $a = 2$ according to Burdine's method [101]. For particular values of the parameters, Equations (5a) and (5b) reduce to the well-known equations proposed by References [101,102]. For two-layered soils, two additional conditions are required at the interface between the two layers:

$$\psi_1(Z_c) = \psi_2(Z_c) = \psi_c,\tag{6a}$$

$$K_1\left[\left(\frac{\partial \psi_1}{\partial z}\right)_{Z_c} - 1\right] = K_2\left[\left(\frac{\partial \psi_2}{\partial z}\right)_{Z_c} - 1\right],\tag{6b}$$

where hereafter the subscripts 1, 2, and c denote variables in the upper layer, in the lower layer, and at the interface, respectively, and Z_c is the interface depth.

In particular cases, it can be useful to express Equation (3) in terms of θ obtaining:

$$\frac{\partial \theta}{\partial t} = \frac{\partial}{\partial z}\left[D(\theta)\frac{\partial \theta}{\partial z} + K(\theta)\right], \tag{7}$$

where $D(\theta)$ is the soil water diffusivity equal to $K(\theta)(d\psi(\theta)/d\theta)$.

3. Point Infiltration Modeling for Homogeneous Soils

Many local infiltration models for vertically homogeneous soils with constant initial soil water content and over horizontal surfaces are widely recognized in the scientific literature [4,5,8–11,17,18, 20,22,23,26,28–30,103]. Furthermore, for isolated storms and when ponding is not achieved instantly, extended forms of the Philip model [45], the Green–Ampt model [104,105], and the Smith and Parlange model [106] have been widely used, whereas for arbitrary rainfall patterns, the model presented in [26] serves as a useful method. These four models, together with a widely adopted equation suggested by References [29,30], have been extensively used in applied hydrology and as building blocks in the development of infiltration approaches at the field scale.

3.1. Horton Empirical Equation

In the empirical equation proposed by References [29,30], the infiltration capacity, f_c, exponentially decreases as follows (see also Figure 2):

$$f_c = f_f + \left(f_0 - f_f\right)exp(-\alpha t), \tag{8}$$

where f_0 and f_f represent the initial and final values of f_c, respectively, and α is the decay constant. When $t\rightarrow\infty$, f_f can be considered equal to the saturated hydraulic conductivity of the soil. If K and D are independent of θ, References [107,108] demonstrated that Equation (8) can be obtained from Equation (7).

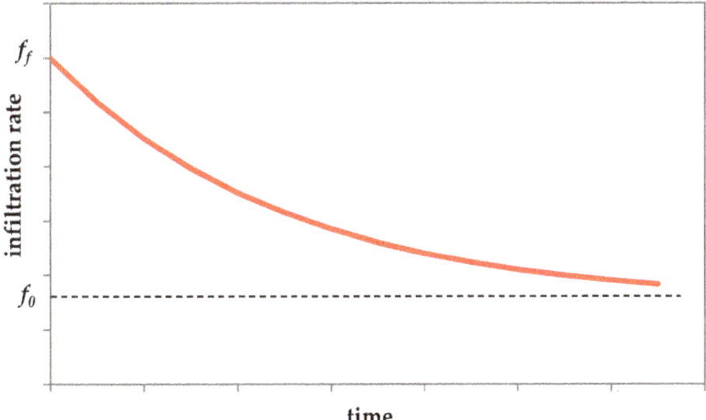

Figure 2. Graphical representation of the Horton empirical equation.

3.2. Philip Equation

A widely adopted analytical solution of the Richards equation was proposed by References [9–11,109] under the conditions of vertically homogeneous soil, constant initial moisture content, and saturated soil surface with immediate ponding. For early to intermediate times the semi-analytical solution for the infiltration capacity, based on a series expansion and truncation after the first two terms, is expressed as:

$$f_c = \frac{1}{2}St^{-1/2} + A \,, \tag{9}$$

where S is the sorptivity, depending on soil properties and initial moisture content, and A is a quantity ranging from 0.38 K_s to 0.66 K_s. For $t \to \infty$, Equation (9) is replaced by $f_c = K_s$. The integration of Equation (9) yields the cumulative infiltration:

$$F = St^{1/2} + At. \tag{10}$$

Philip's model was extended for applications to less restrictive conditions. For a constant rainfall rate $r > K_s$, surface saturation occurs at a time $t_p > 0$ and, following [45], infiltration can be described through an equivalent time origin, t_0, for potential infiltration after ponding as:

$$t_p = \frac{S^2 \left(r - \frac{A}{2} \right)}{2r(r - A)^2} \,, \tag{11}$$

$$t_0 = t_p - \frac{1}{4A^2} \left[\left(S^2 + 4AF_p \right)^{1/2} - S \right]^2 \,, \tag{12}$$

$$f_c = \frac{1}{2}S(t - t_0)^{-1/2} + A, \ t > t_p \,. \tag{13}$$

For unsteady rainfall under the condition of a continuously saturated surface for $t > t_p$, the infiltration process can be represented adopting a similar procedure. Generally, S and A are derived from the calibration of hydrological models; however, S can also be approximated as [110]:

$$S = [2(\phi - \theta_i)K_s|\psi_{av}|]^{1/2}, \tag{14}$$

where ϕ is the soil porosity and ψ_{av} is the soil water matric capillary head at the wetting front.

3.3. Green–Ampt Model

This model represents infiltration into homogeneous soils under the conditions of continuously saturated soil surface and uniform initial soil moisture as:

$$f_c = K_s \left[1 - \frac{\psi_{av}(\theta_s - \theta_i)}{F} \right] , \tag{15}$$

where F the cumulative depth of infiltrated water. To express the infiltration as a function of time, this equation can be solved after the substitution $f_c = dF/dt$. The resulting equation [45] is:

$$F = K_st - \psi_{av}(\theta_s - \theta_i)ln\left[1 - \frac{F}{\psi_{av}(\theta_s - \theta_i)} \right] . \tag{16}$$

Equation (16) assumes immediate ponding and an infinite supply of water at the surface. Under more general conditions, with a constant rainfall rate $r > K_s$, that begins at the time $t = 0$, surface saturation is reached at a time $t_p > 0$. For $t \le t_p$, the infiltration rate q_0 is equal to r and later to the infiltration capacity. Mein and Larson [104] formulated this process through Equation (15) as:

$$r - K_s = -\frac{K_s\psi_{av}(\theta_s - \theta_i)}{\int_0^{t_p} r \, dt} \,, \tag{17}$$

which leads to the determination of t_p as:

$$t_p = -\frac{K_s \psi_{av}(\theta_s - \theta_i)}{r(r - K_s)} .$$ (18)

Furthermore, for $t > t_p$, Equation (16) becomes:

$$F = F_p - \psi_{av}(\theta_s - \theta_i) ln\left[\frac{F - \psi_{av}(\theta_s - \theta_i)}{F_p - \psi_{av}(\theta_s - \theta_i)}\right] + K_s(t - t_p)$$ (19)

Equation (19) can be solved at each time, for example, by successive substitutions of F which is then substituted into Equation (15) to obtain the corresponding value of f_c.

3.4. Parlange–Lisle–Braddock–Smith Model

A 3-parameter model obtained through an analytical integration of the Richards equation has been formulated by Reference [106] and can be expressed as:

$$f_c = K_s\left[1 + \frac{\alpha}{exp\left(\frac{\alpha F'}{G(\theta_s - \theta_i)}\right) - 1}\right],$$ (20)

where $F' = F - K_i t$ is the cumulative dynamic infiltration rate, α is a parameter linked to the behavior of hydraulic conductivity and soil water diffusivity as functions of θ, and G is the integral capillary drive defined by:

$$G = \frac{1}{K_s} \int_{\theta_i}^{\theta_s} D(\theta)d\theta$$ (21)

Equation (20) includes as limiting forms [60] the Reference [17] equation ($\alpha = 1$) and the Green–Ampt equation ($\alpha \to 0$). It can be applied to determine t_p and f_c for any rainfall pattern, and for $t > t_p$ can be rewritten under the condition of surface saturation in a time dependent form as:

$$[(1 - \alpha)K_s - K_i](t - t_p) = F' - F'_p - \frac{(\theta_s - \theta_i)K_s G}{K_d} ln\left\{\frac{exp\left[\frac{\alpha F'}{(\theta_s - \theta_i)G}\right] - 1 + \frac{\alpha K_s}{K_d}}{exp\left[\frac{\alpha F'_p}{(\theta_s - \theta_i)G}\right] - 1 + \frac{\alpha K_s}{K_d}}\right\},$$ (22)

where $K_d = K_s - K_i$ and $F'_p = F'(t_p)$. The quantities F'_p and t_p are the values of F' and t at ponding, respectively, at which Equation (20) with $f_c = r(t_p)$ is first satisfied. The value of α usually ranges from 0.8 to 0.85 [3].

3.5. Corradini–Melone–Smith Semi-Analytical/Conceptual Model

When the hypothesis of uniform initial soil moisture cannot be guaranteed, the models presented in the previous subsections cannot be applied. For complex rainfall patterns involving rainfall hiatus periods or a rainfall rate after time to ponding less than soil infiltration capacity, an approach for the application of the aforementioned classical models was developed [111–113] starting from the time compression approximation proposed by Reference [114] for post-hiatus rainfall producing immediate ponding (see also [1]). However, by comparing results with the Richards equation, Reference [22] showed that this approach was not sufficiently accurate because it neglects the soil water redistribution process (Figure 3) which is particularly important when long periods with a light rainfall or a rainfall hiatus occur.

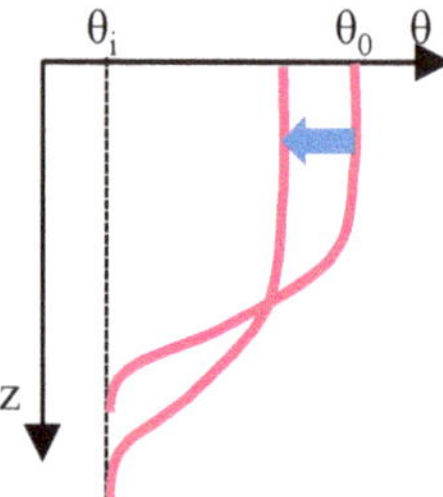

Figure 3. Graphical representation of the soil water redistribution process.

Models combining infiltration and redistribution are therefore the best solution when soils are subjected to complex rainfall patterns, which are prevalent under natural conditions (see also [115,116]). Dagan and Bresler [20] developed an analytical model along this line starting from depth integrated forms of Darcy's law and the continuity equation and using simplifications in the initial and surface boundary conditions that make easier areal investigations but reduce practical applications at the local scale. In any case, the model is not applicable to local studies with erratic rainfalls producing successive infiltration–redistribution cycles. A more general model was formulated by Reference [26] starting from the same integrated equations, and then combined with a conceptual representation of the wetting soil moisture profile.

To demonstrate the structure of the model presented in [26], a specific rainfall pattern to describe all the involved components is presented here. The application to erratic rainfall is then a straightforward extension. Let us consider a stepwise rainfall pattern involving successive periods of rainfall with constant $r > K_s$, separated by periods with $r = 0$ (see Figure 4). We denote by t_1 the duration of the first pulse, t_2 and t_3, the beginning and end of the second pulse, respectively, and t_4 the beginning of the third pulse. The model was derived considering a soil with a constant value of θ_i and combining the depth-integrated forms of Darcy's law and the continuity equation. In addition, as the event progresses in time, a dynamic wetting profile, of lowest depth Z and represented by a distorted rectangle through a shape factor $\beta(\theta_0) \leq 1$, was assumed. The resulting ordinary differential equation is:

$$\frac{d\theta_0}{dt} = \frac{(\theta_0 - \theta_i)\beta(\theta_0)}{F'\left[(\theta_0 - \theta_i)\frac{d\beta(\theta_0)}{d\theta_0} + \beta(\theta_0)\right]}\left[q_0 - K_0 - \frac{(\theta_0 - \theta_i)G(\theta_i,\theta_0)\beta(\theta_0)pK_0}{F'}\right], \tag{23}$$

where p is a parameter linked with the profile shape of θ and $G(\theta_i,\theta_0)$ is expressed by Equation (21) modified by the substitution of θ_s with θ_0 and K_s with K_0. Equation (23) can be applied for $0 < t < t_2$, and the profile shape of $\theta(z)$ is approximated [23] by:

$$\frac{\theta(z) - \theta_i}{\theta_0 - \theta_i} = 1 - exp\left[\frac{\beta z(\theta_0 - \theta_i) - F'}{(\beta - \beta^2) - F'}\right]. \tag{24}$$

Functional forms for β and p were obtained by calibration using results provided by the Richards equation applied to a silty loam soil, specifically:

$$\beta(\theta_0) = 0.6\frac{\theta_s - \theta_i}{\theta_s - \theta_r} + 0.4, \tag{25a}$$

$$\beta p = 0.98 - 0.87\,exp\left(-\frac{r}{K_s}\right) \qquad \frac{d\theta_0}{dt} \geq 0, \tag{25b}$$

$$\beta p = 1.7 \qquad \frac{d\theta_0}{dt} < 0. \tag{25c}$$

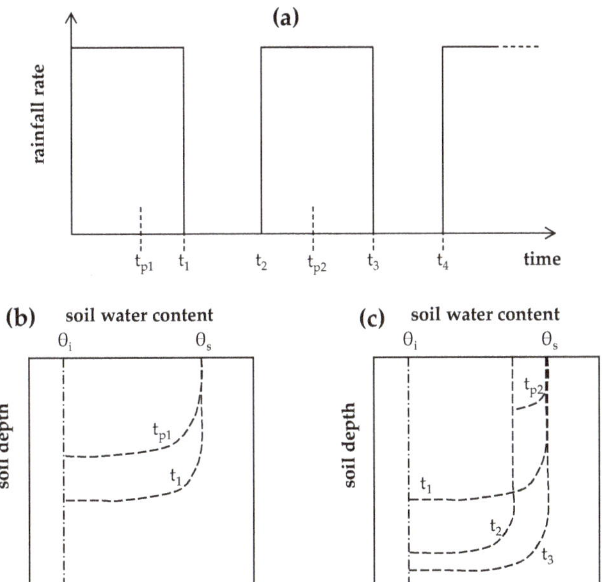

Figure 4. (**a**) Rainfall pattern selected to describe the Corradini–Melone–Smith semi-analytical model for point infiltration. (**b**,**c**) Profiles of soil water content at various times indicated in (**a**) and associated with different infiltration-redistribution stages. For symbols, see the text.

Equation (23) can be solved numerically. For $q_0 = r$, with $F' = (r-K_i)t$, it gives $\theta_0(t)$ until time to ponding, t'_p, corresponding to $\theta_0 = \theta_s$ and $d\theta_0/dt = 0$. Then, for $t'_p < t \leq t_1$, with $\theta_0 = \theta_s$ and $d\theta_0/dt = 0$, it provides the infiltration capacity ($q_0 = f_c$) and for $t_1 < t < t_2$, with $q_0 = 0$, it gives $d\theta_0/dt < 0$ thus describing the redistribution process.

The second rainfall pulse leads to a new time to ponding, t''_p, but reinfiltration occurs according to two alternative approaches determined by a comparison of r and the downward redistribution rate, $D_F(t = t_2)$, expressed by:

$$D_F = \left(-\frac{1}{\beta}\frac{d\beta}{d\theta_0} - \frac{1}{\theta_0 - \theta_i} \right) F'\frac{d\theta_0}{dt} - K_i \quad t = t_2. \tag{26}$$

More specifically, for $r \leq D_F$, the reinfiltrated water is distributed to the whole dynamic profile and $\theta_0(t)$ can be still computed by Equation (23), whereas for $r > D_F$, the profile of $\theta(z, t = t_2)$ is assumed temporarily invariant and starts a superimposed secondary wetting profile which advances alongside the pre-existing profile according to Equation (23) modified by substituting θ_i with $\theta_0(t_2)$ and F' with F'_{2t} accumulated for $t \geq t_2$. If the secondary profile reaches the depth of the first steady one, the compound profile reduces to a single profile and then Equation (23) can be again applied (see Figure 3 for a schematic representation of $\theta(z,t)$). On the other hand, if at $t = t_3$ the secondary profile has not caught up with the first one, redistribution is first applied to the secondary profile and then re-established to the single profile in the successive rainfall hiatus. Finally, in the case at $t = t_4$, the $\theta(z)$ profile is still compound and r is larger than $D_F(t_4)$, and a procedure of consolidation that merges the composite profile is applied early to avoid the formation of a further additional profile.

This model incorporates all the components required for application to any natural rainfall pattern. It was calibrated by Reference [26] for a silty loam soil, and then tested using different soils from clay loam to sandy loam soil types. Weighted implicit finite difference solutions of the Richards equation were used as a benchmark. For each soil, the model accuracy was found to be acceptable in terms of both infiltration rate and soil water content, even though better results were obtained for fine-textured soils. In addition, the model was found to simulate fairly well the $\theta(z,t)$ profiles observed in laboratory (Figure 5) and field experiments [117,118].

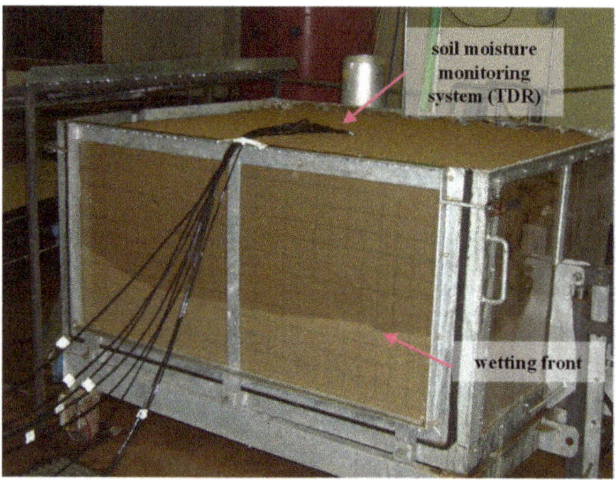

Figure 5. Laboratory experimental system adopted by References [117,118] to verify the Corradini–Melone–Smith model.

4. Point Infiltration Modeling for Vertically Non-Uniform Soils

A two-layer approximation, with each layer being schematized as homogeneous, is frequently used to set up models of infiltration for natural soils. The process of formation of a sealing layer was accurately examined by References [31,119], and that of disruption was considered by References [120–122]. Evidence of the role of crusted soils in semi-arid regions has been recently provided by Reference [123]. Green–Ampt-based models for infiltration into stable crusted soils were proposed by References [33,34,36,39]. An efficient approach that represents transient infiltration into crusted soils was proposed by Reference [40], whereas Reference [41] formulated a model that describes upper-layer dynamics for a ponded upper boundary. On the other hand, vertical profiles with a more permeable upper layer are observed in hydrological practice and can be also used, for example, as a first approximation in the representation of infiltration into homogeneous soils with grassy vegetation [124]. In the latter layering type, the simple model presented by Reference [45] can be usefully applied for infiltration into a saturated soil surface. Under more general conditions, the model by Reference [43] appears to be accurate with modest computational effort.

4.1. Green–Ampt-Based Model for a Layered Soil

A model for infiltration into a two-layered soil with a more permeable upper layer under the condition of continuously saturated soil surface has been formulated by Reference [45]. The classical Green–Ampt equation is applied until the wetting front is in the upper layer, then the following equations are used:

$$L_2 \frac{(\theta_{2s} - \theta_{2i})}{K_{2s}} + \frac{[(\theta_{2s} - \theta_{2i})Z_c K_{2s} - (\theta_{2s} - \theta_{2i})K_{1s}(\psi_{av\,2} + Z_c)]}{K_{1s}K_{2s}} ln\left[1 + \frac{L_2}{\psi_{av\,2} + Z_c}\right] = t\,, \tag{27}$$

$$F = Z_c(\theta_{1s} - \theta_{1i}) + L_2(\theta_{2s} - \theta_{2i})\,, \tag{28}$$

$$f_c = \frac{K_{1s}K_{2s}}{Z_c K_{2s} + L_2 K_{1s}}(\psi_{av\,2} + Z_c + L_2)\,, \tag{29}$$

where L_2 is the depth of the wetting front below the interface. Equation (27) can be solved at each time for L_2 by successive substitutions; L_2 is then used in Equations (28) and (29) to determine F and f_c, respectively.

4.2. Corradini–Melone–Smith Semi-Analytical/Conceptual Model for a Two-Layered Soil

A semi-analytical/conceptual model applicable to any horizontal two-layered soil where either layer may be more permeable has been formulated by Reference [43]. It relies on the same elements previously used by Reference [26] but adopted here in each layer and integrated at the interface between the two layers by the boundary conditions expressing continuity of flow rate and capillary head (Equations (6a) and (6b)). In addition, at the lower boundary for $t > 0$, we have $\psi_2 = \psi_i$. The initial condition is $\psi_1 = \psi_2 = \psi_i$ constant at $t = 0$, and at the interface $q(Z_c)$ is approximated through the downward flux in the upper layer as:

$$q(Z_c) = K_{1s}\frac{G_1(\psi_c, \psi_{10})}{Z_c} + K_{1c} \, , \tag{30}$$

where $G(\psi_c, \psi_{10})$ is expressed by Equation (21) modified by the substitutions of $D(\theta)d\theta$ with $K(\psi)d\psi$, θ_s with ψ_{10}, and θ_i with ψ_c. As long as water does not infiltrate in the lower layer, the model presented in [26] is used, then starting from the time t_c when the wetting front enters the lower layer, the following system of two ordinary differential equations may be applied:

$$\frac{d\psi_{10}}{dt} = \frac{1}{\gamma Z_c C_1(\psi_{10})}\left[q_{10} - K_{1c} - \frac{K_{1s}G_1(\psi_c, \psi_{10})}{Z_c}\right] - \frac{(1-\gamma)C_1(\psi_c)}{\gamma C_1(\psi_{10})}\frac{d\psi_c}{dt} \, t \geq t_c \, , \tag{31}$$

$$\frac{d\psi_c}{dt} = \frac{1}{P_L(\psi_c, t)}\left[K_{1c} + \frac{K_{1s}G_1(\psi_c, \psi_{10})}{Z_c} - K_{2c} - \frac{\beta_2(\theta_{2c})p_2(\theta_{2c} - \theta_{2i})K_{2s}G_2(\psi_i, \psi_c)}{F_2}\right] \, t \geq t_c, \tag{32}$$

with $P_L(\psi_c, t)$ and F_2 defined as:

$$P_L(\psi_c, t) = \left[\beta_2(\theta_{2c}) + \frac{d\beta_2}{d\theta_{2c}}(\theta_{2c} - \theta_{2i})\right]\frac{F_2}{(\theta_{2c} - \theta_{2i})\beta_2(\theta_{2c})}C_2(\psi_c) \, , \tag{33}$$

$$F_2 = F - Z_c[\gamma(\theta_{1s} - \theta_{1i}) + (1 - \gamma)(\theta_{1c} - \theta_{1i})] - K_{2i}t, \tag{34}$$

and $C_1(\psi_{10}) = d\theta_{10}/d\psi_{10}$, $C_1(\psi_c) = d\theta_{1c}/d\psi_{1c}$. The quantity γ represents a conceptual proportion of the upper layer where θ is increasing due to rainfall and is assumed equal to 0.85 [42], β_2 and p_2 are determined by Equations (25a–c) but applied using θ_{20}, θ_{2s}, θ_{2i}, and θ_{2r} and substituting r with $q(Z_c)$. On the basis of the same stepwise rainfall pattern earlier adopted to explain the model presented in [26], Equations (31) and (32) may be used for $t_c < t < t_2$. Then, the two-layer model has to be applied in each layer by analogy with the procedure described for homogeneous soils; in particular, compound and consolidated profiles develop in each layer. In the underlying soil, the generation of additional profiles occurs through $q(Z_c)$ in substitution of r. On the basis of the described steps, model application to arbitrary rainfall patterns is straightforward. The solution of the above system, Equations (31) and (32), may be obtained by a library routine for the Runge–Kutta–Verner fifth-order method with a variable time step. Calibration and testing of the model were performed through a comparison with numerical solutions of the Richards equation. Three soils (clay loam, silty loam, and sandy loam) with a variety of thicknesses were combined to realize two-layered soils, where either layer was more permeable, that were selected as test cases. In all instances, the simulations involved the cycle of infiltration–redistribution–reinfiltration. The infiltration rate as well as the water content at the surface and interface were found to be very accurately estimated by the semi-analytical/conceptual model.

5. Areal Infiltration Models over Soil with Variable Hydraulic Properties

At the field scale, considering the significant spatial variability of the main soil hydraulic properties, rainfall infiltration modeling is not analytically tractable, whereas the use of accurate Monte Carlo (M C) simulation techniques imposes an enormous computational burden for routine applications.

In the Introduction section, the evolution of studies for field-scale infiltration models for spatial variability in soil hydraulic properties and rainfall was presented. Models for infiltration at the field scale have been recently developed and represent a useful support for practical hydrological purposes. Two models characterized by significant differences in complexity and application area are presented here.

5.1. Smith and Goodrich Approach

A semi-empirical model to determine the areal-average infiltration rate into areas with random spatial variability of K_s has been proposed by Reference [60]. The authors assumed a lognormal probability density function (PDF) of K_s with a mean value <K_s> and a coefficient of variation $CV(K_s)$, and considered one realization of the random variable. Then, adopting the model presented in [106] (see also [22]) and the Latin Hypercube sampling method, and through a large number of simulations performed for many values of $CV(K_s)$ and rainfall rates, they developed the following effective areal relation for the scaled areal-average infiltration rate, I_e^*, linked with the corresponding scaled cumulative depth, F_e^*:

$$I_e^* = 1 + (r_e^* - 1)\left\{1 + \left[\frac{r_e^* - 1}{\alpha}\left(e^{\alpha F_e^*} - 1\right)\right]^{c_a}\right\}^{-1/c_a} \qquad r_e^* > 1, \tag{35}$$

with

$$c_a \cong 1 + \frac{0.8}{[CV(K_s)]^{1.3}}\left[1 - e^{(0.85(r_b^* - 1))}\right], \tag{36}$$

where $I_e^* = I_e/K_e$, $F_e^* = F_e/[G(\theta_s - \theta_i)]$, $r_e^* = r/K_e$ and, $r_b^* = r/<K_s>$. The quantity K_e denotes the areal effective value of K_s given by:

$$K_e = \int_0^r K f_{K_s}(K)dK + [1 - \int_0^r f_{K_s}(K)dK]r, \tag{37}$$

where $f_{K_s}(K)$ is the PDF of K_s. Finally, Equation (35) may be also applied for r variable with time.

5.2. Govindaraju–Corradini–Morbidelli Semi-Analytical/Conceptual Model

Govindaraju et al. [72] formulated a semi-analytical model to estimate the expected field-scale infiltration rate <$I_n(F)$> under the condition of negligible effects of the run-on process. The model incorporates heterogeneity of both K_s and r assumed as random variables with a lognormal and a uniform PDF, respectively. The quantity <$I_n(F)$> is estimated through the averaging procedure over the ensemble of two-dimensional realizations of K_s and r. For the sake of simplicity, we first examine the model under a steady-rainfall condition, and then provide the guidelines for applications to a time-varying rainfall rate.

Starting from the extended Green–Ampt model and choosing F as the independent variable, <$I_n(F)$> at a given F can be written as:

$$\langle I(F)\rangle = \int_0^\infty \int_{K_c}^\infty r f_r(r) f_{K_s}(K) dr\, dK + \int_0^\infty \int_0^{K_c}\left(1 + \frac{\psi(\theta_s - \theta_i)}{F}\right) K f_r(r) f_{K_s}(K) dr\, dK, \tag{38}$$

where $f_r(r)$ and $f_{K_s}(K)$ are the PDFs of r and K_s, respectively, with K_c which denotes the maximum value of K_s leading to surface saturation in the i-th cell, determined by:

$$K_c = \frac{Fr_i}{\psi(\theta_s - \theta_i) + F} = F_c r_i \, . \tag{39}$$

Equation (38) may be expressed as:

$$
\begin{aligned}
\langle I_n(F) \rangle = {} & \frac{1}{2RF_c^2}\{M_{K_s}[(r_{min}+R)F_c,\, 2] - M_{K_s}[(r_{min})F_c,\, 2]\} \\
& - \frac{r_{min}^2}{2R}\{M_{K_s}[(r_{min}+R)F_c,\, 0] - M_{K_s}[(r_{min})F_c,\, 0]\} \\
& + \left(r_{min} + \frac{R}{2}\right)\{1 - M_{K_s}[(r_{min}+R)F_c,\, 0]\} \\
& + \frac{1}{F_c}\left(\frac{r_{min}+R}{R}\right)\{M_{K_s}[(r_{min}+R)F_c,\, 1] - M_{K_s}[(r_{min})F_c,\, 1]\} \\
& - \frac{1}{F_c}\left(\frac{1}{RF_c}\right)\{M_{K_s}[(r_{min}+R)F_c,\, 2] - M_{K_s}[(r_{min})F_c,\, 2]\} \\
& + \frac{1}{F_c}\{M_{K_s}[(r_{min})F_c,\, 1]\}\, ,
\end{aligned}
\tag{40}
$$

with r_{min} and $r_{min} + R$ extreme values of the PDF of r and M_{K_s} given by:

$$M_{K_s}(K_a, \omega) = \int_0^{K_a} K^\omega f_K(K)dK\, , \tag{41}$$

where K_a and ω stand for the first and the second argument, respectively, of the M_{K_s} function. To relate time to F, an implicit relation between the expected value of t, $<t>$, and F is provided as:

$$
\begin{aligned}
\langle t(F) \rangle = {} & \frac{F}{\langle r \rangle}\{1 - M_{K_s}[\langle r \rangle F_c, 0]\} + \left[F + \psi(\theta_s - \theta_i)ln\left(\frac{\psi(\theta_s-\theta_i)}{\psi(\theta_s-\theta_i)+F}\right)\right] \\
& \{M_{K_s}[\langle r \rangle F_c, -1]\} + \psi(\theta_s - \theta_i)\sum_{j=1}^{\infty}\frac{1}{(j+1)\langle r \rangle^{\,j+1}}\{M_{K_s}[\langle r \rangle F_c, j]\}\, .
\end{aligned}
\tag{42}
$$

To extend the model by incorporating the run-on effect, an additional empirical term is used in the form presented in [73]:

$$\langle I(t) \rangle \cong \langle I_n(t) \rangle + \langle r \rangle\, a \left(\frac{t}{t_{pa}}\right)^b exp\left(-c\frac{t}{t_{pa}}\right), \tag{43}$$

where t_{pa} is the time to ponding (see Equation (18)) associated with $<r>$ and $<K_s>$. The parameters a, b, and c are expressed by:

$$a = 2.8[CV(r) + CV(K_s)]^{0.36}, \tag{44}$$

$$b = 5.35 - 6.32[CV(r)CV(K_s)], \tag{45}$$

$$c = 2.7 + 0.3\left[\frac{\langle r \rangle / \langle K_s \rangle}{CV(r)CV(K_s)}\right]^{0.3}, \tag{46}$$

where Equation (44) holds for $\theta_i \ll \theta_s$ and for $\theta_i \rightarrow \theta_s$ we have $a \rightarrow 0$. Furthermore, Equation (46) is undefined for $CV(r)$ and/or $CV(K_s)$ equal to 0. In Equations (43)–(46), length units are in mm and time is expressed in hours.

When the spatial heterogeneity of r can be neglected, with spatially uniform and steady rainfall and negligible run-on, Equations (40) and (42) can be replaced by [61]:

$$\langle I(F) \rangle = r[1 - M_{K_s}(K_c, 0)] + \frac{\psi(\theta_s - \theta_i) + F}{F}M_{K_s}(K_c, 1)\, , \tag{47}$$

$$\langle t \rangle = \frac{F}{r}\{1 - M_{K_s}[K_c, 0]\} + \left[F + \psi(\theta_s - \theta_i)ln\left(\frac{\psi(\theta_s - \theta_i)}{\psi(\theta_s - \theta_i) + F}\right)\right]$$

$$\{M_{K_s}[K_c, -1]\} + \psi(\theta_s - \theta_i)\sum_{i=1}^{\infty}\frac{M_{K_s}[K_c, i]}{(i+1)r^{i+1}}\,. \tag{48}$$

The last formulation was also extended for applications involving r variable in time. The same method can be used to adapt Equations (40) and (42) for unsteady rainfall patterns. Furthermore, the additional empirical term for run-on could be adapted for unsteady rainfalls following the guidelines indicated by Reference [73].

The solution of the model even in the conditions of coupled spatial variability of r and K_s is fairly simple and requires limited computational effort.

The model for coupled heterogeneity of r and K_s (Equations (40), (42), and (43)) was validated by comparison with the results derived starting from MC sampling and using a combination of the extended Green–Ampt formulation at the local scale with the kinematic wave approximation [125] that is required to represent run-on. Through a wide variety of simulations it was shown that: (1) the model produced very accurate estimates of <*I*> over a clay loam soil and a sandy loam soil; (2) the spatial heterogeneity of both r and K_s can be neglected only when <*r*> $\gg$ <*K*> or for storm durations much greater than t_{pa}; (3) the effects on <*I*> produced by significant values of $CV(K_s)$ and $CV(r)$ are similar; (4) run-on plays a significant role for moderate storms and high values of $CV(K_s)$ and $CV(r)$; and (5) the model can be simplified using Equations (47) and (48) for $CV(r)$ substantially less than $CV(K_s)$ and steady rainfalls.

6. Conclusions and Open Problems

In spite of the continuous developments in infiltration modeling, challenges regarding infiltration exist for many scales of hydrologic interest. The conflicting results from different studies on the effect of slope on infiltration over homogeneous surfaces show that a solution continues to elude researchers. Careful experimental and theoretical investigations in this regard are needed to fully comprehend this fundamental infiltration behavior over sloping surfaces.

The estimate of infiltration at different spatial scales (i.e., from the local to watershed scales) is a complex problem as further challenges are imposed by the natural spatial variability of soil hydraulic characteristics and that of rainfall. An important issue to be addressed when areal estimates are involved is that concerning the determination of <K_s>, $CV(K_s)$, <*r*>, and $CV(r)$ together with the corresponding quantities for soil moisture content [126].

The models presented here apply to infiltration into a soil matrix. When macropore flow is significant, the problem becomes much more complicated even though simplified approaches have been proposed. For example, two practical approximations to describe infiltration into soils with macropores rely upon the use of modified values of the saturated hydraulic conductivity of the soil matrix [127] or upon the representation of the two processes of infiltration controlled by the matrix potential and the macropore volume [128]. However, there exist many other models to represent the macropore flow [129], based on the single continuum approach (e.g., [130]), the dual continuum approach, the dual permeability approach (e.g., [131]), and the dual porosity approach (e.g., [132]). Notwithstanding the high number of simplified models, and given the difficulty to investigate the Navier–Stokes equations over finite soil portions, macropore flow still needs a convincing physical theory for the scales of practical interest.

Finally, all these models are formulated for horizontal land surfaces. Extensions of the classical infiltration theory to inclined surfaces were proposed by References [81,87]; however, these theories do not explain the results of laboratory experiments, for example those performed on bare soils in the absence of erosion and a sealing layer [88,92]. The modeling of the slope effects has to be therefore considered as an open problem [97], in particular when surfaces with vegetation are involved [95].

Further challenges exist in our ability to independently measure soil properties at the local scale to identify the true nature of field-scale variability. The use of common measuring instruments for soil hydraulic properties does not yield consistent estimates of variability (e.g., [133]). Both the inter- and intra-instrument errors contribute to a level of uncertainty that has not been understood. Even though measurement techniques are not the topic of this review, measurement errors and uncertainty nevertheless influence modeling efforts that rely on these data. Efforts to appropriately combine disparate measurement results are also needed to realize the full worth of the data that are generated from expensive and time-consuming experimental campaigns.

Field-scale experiments measuring runoff and deep flow are influenced by spatial variability as well, but as yet no theory exists for elucidating the underlying spatial variability from these experiments. Current approaches (e.g., [134]) rely on brute force calibration techniques; however, such methods are often plagued by identifiability and non-uniqueness problems. Moreover, calibration is known to compensate for various model deficiencies, and authors deriving parameter estimates from these approaches must be cognizant of the role played by the underlying model structure. Better conceptual and theoretical underpinnings are needed to move the science of infiltration forward.

Author Contributions: Investigation, Writing—original draft preparation and Writing—review and editing, R.M., C.C., C.S., A.F., J.D., R.S.G.

Funding: This research was partly financed by the Italian Ministry of Education, University and Research (PRIN 2015).

Conflicts of Interest: The authors declare no conflict of interest.

References

1. Brutsaert, W. *Hydrology—An Introduction*; Cambridge University Press: Cambridge, UK, 2005; ISBN 978-0521824798.
2. Assouline, S. Infiltration into soils: Conceptual approaches and solutions. *Water Resour. Res.* **2013**, *49*, 1755–1772. [CrossRef]
3. Smith, R.E. *Infiltration Theory for Hydrologic Applications*; Water Resources Monograph; American Geophysical Union: Washington, DC, USA, 2002; Volume 15, ISBN 9780875903194.
4. Green, W.A.; Ampt, G.A. Studies on soil physics: 1. The flow of air and water through soils. *J. Agric. Sci.* **1911**, *4*, 1–24. [CrossRef]
5. Kostiakov, A.N. On the dynamics of the coefficient of water-percolation in soils and on the necessity for studying it from a dynamic point of view for purposes of amelioration. *Trans. Sixth Comm. Int. Soc. Soil Sci.* **1932**, *1 Pt A*, 17–21.
6. Horton, R.E. An approach toward a physical interpretation of infiltration-capacity. *Soil Sci. Soc. Am. J.* **1940**, *5*, 399–417. [CrossRef]
7. Holtan, H.N. *A Concept for Infiltration Estimates in Watershed Engineering*; USDA Bulletin: Washington, DC, USA, 1961; pp. 41–51.
8. Swartzendruber, D. A quasi solution of Richard equation for downward infiltration of water into soil. *Water Resour. Res.* **1987**, *5*, 809–817. [CrossRef]
9. Philip, J.R. The theory of infiltration: 1. The infiltration equation and its solution. *Soil Sci.* **1957**, *83*, 345–357. [CrossRef]
10. Philip, J.R. The theory of infiltration: 2. The profile at infinity. *Soil Sci.* **1957**, *83*, 435–448. [CrossRef]
11. Philip, J.R. The theory of infiltration: 4. Sorptivity algebraic infiltration equation. *Soil Sci.* **1957**, *84*, 257–264. [CrossRef]
12. Parlange, J.-Y. Theory of water-movement in soils: 2. One-dimensional infiltration. *Soil Sci.* **1971**, *111*, 170–174. [CrossRef]
13. Parlange, J.-Y. Theory of water-movement in soils: 8. One-dimensional infiltration with constant flux at surface. *Soil Sci.* **1972**, *114*, 1–4. [CrossRef]
14. Soil Conservation Service (SCS). *National Engineering Handbook*; Section 4, Hydrology; USDA: Washington, DC, USA, 1972.
15. Brutsaert, W. The concise formulation of diffusive sorption of water in a dry soil. *Water Resour. Res.* **1976**, *12*, 1118–1124. [CrossRef]
16. Brutsaert, W. Vertical infiltration in dry soils. *Water Resour. Res.* **1977**, *13*, 363–368. [CrossRef]
17. Smith, R.E.; Parlange, J.-Y. A parameter-efficient hydrologic infiltration model. *Water Resour. Res.* **1978**, *14*, 533–538. [CrossRef]

18. Broadbridge, P.; White, I. Constant rate rainfall infiltration: A versatile nonlinear model. 1. Analytic solution. *Water Resour. Res.* **1988**, *24*, 145–154. [CrossRef]

19. White, I.; Broadbridge, P. Constant rate rainfall infiltration: A versatile nonlinear model. 2. Application of solutions. *Water Resour. Res.* **1988**, *24*, 155–162. [CrossRef]

20. Dagan, G.; Bresler, E. Unsaturated flow in spatially variable fields. 1. Derivation of models of infiltration and redistribution. *Water Resour. Res.* **1983**, *19*, 413–420. [CrossRef]

21. Haverkamp, R.; Parlange, J.-Y.; Starr, J.L.; Schmitz, G.; Fuentes, C. Infiltration under ponded conditions: 3. A predictive equation based on physical parameters. *Soil Sci.* **1990**, *149*, 292–300. [CrossRef]

22. Smith, R.E.; Corradini, C.; Melone, F. Modeling infiltration for multistorm runoff events. *Water Resour. Res.* **1993**, *29*, 133–144. [CrossRef]

23. Corradini, C.; Melone, F.; Smith, R.E. Modeling infiltration during complex rainfall sequences. *Water Resour. Res.* **1994**, *30*, 2777–2784. [CrossRef]

24. Ross, P.J.; Parlange, J.-Y. Comparing exact and numerical solutions of Richards equation for one-dimensional infiltration and drainage. *Soil Sci.* **1994**, *157*, 341–344. [CrossRef]

25. Barry, D.A.; Parlange, J.-Y.; Haverkamp, R.; Ross, P.J. Infiltration under ponded conditions: 4. An explicit predictive infiltration formula. *Soil Sci.* **1995**, *160*, 8–17. [CrossRef]

26. Corradini, C.; Melone, F.; Smith, R.E. A unified model for infiltration and redistribution during complex rainfall patterns. *J. Hydrol.* **1997**, *192*, 104–124. [CrossRef]

27. Furman, A.; Warrick, A.W.; Zerihun, D.; Sanchez, C.A. Modified Kostiakov infiltration function: Accounting for initial and boundary conditions. *J. Irrig. Drain. Eng.* **2006**, *132*, 587–596. [CrossRef]

28. Wang, K.; Yang, X.; Liu, X.; Liu, C. A simple analytical infiltration model for short-duration rainfall. *J. Hydrol.* **2017**, *555*, 141–154. [CrossRef]

29. Horton, R.E. The role of infiltration in the hydrological cycle. *Trans. Am. Geophys. Union* **1933**, *14*, 446–460. [CrossRef]

30. Horton, R.E. Analysis of runoff plat experiments with varying infiltration capacity. *Trans. Am. Geophys. Union* **1939**, *20*, 693–711. [CrossRef]

31. Mualem, Y.; Assouline, S.; Eltahan, D. Effect of rainfall-induced soil seals on soil water regime: Wetting processes. *Water Resour. Res.* **1993**, *29*, 1651–1659. [CrossRef]

32. Taha, A.; Gresillon, J.M.; Clothier, B.E. Modelling the link between hillslope water movement and stream flow: Application to a small Mediterranean forest watershed. *J. Hydrol.* **1997**, *203*, 11–20. [CrossRef]

33. Hillel, D.; Gardner, W.R. Transient infiltration into crust-topped profile. *Soil Sci.* **1970**, *109*, 64–76. [CrossRef]

34. Ahuja, L.R. Modeling infiltration into surface-sealed soils by the Green-Ampt approach. *Soil Sci. Soc. Am. J.* **1983**, *47*, 412–418. [CrossRef]

35. Beven, K. Infiltration into a class of vertical non-uniform soils. *Hydrol. Sci. J.* **1984**, *29*, 425–434. [CrossRef]

36. Vandervaere, J.-P.; Vauclin, M.; Haverkamp, R.; Peugeot, C.; Thony, J.-L.; Gilfedder, M. Prediction of crust-induced surface runoff with disc infiltrometer data. *Soil Sci.* **1998**, *163*, 9–21. [CrossRef]

37. Selker, J.S.; Duan, J.F.; Parlange, J.-Y. Green and Ampt infiltration into soils of variable pore size with depth. *Water Resour. Res.* **1999**, *35*, 1685–1688. [CrossRef]

38. Chu, X.; Marino, M.A. Determination of ponding condition and infiltration in layered soils under unsteady rainfall. *J. Hydrol.* **2005**, *313*, 195–207. [CrossRef]

39. Moore, I.D. Infiltration equations modified for surface effects. *J. Irrig. Drain. Eng.* **1981**, *107*, 71–86.

40. Smith, R.E. Analysis of infiltration through a two-layer soil profile. *Soil Sci. Soc. Am. J.* **1990**, *54*, 1219–1227. [CrossRef]

41. Philip, J.R. Infiltration into crusted soils. *Water Resour. Res.* **1998**, *34*, 1919–1927. [CrossRef]

42. Smith, R.E.; Corradini, C.; Melone, F. A conceptual model for infiltration and redistribution in crusted soils. *Water Resour. Res.* **1999**, *35*, 1385–1393. [CrossRef]

43. Corradini, C.; Melone, F.; Smith, R.E. Modeling local infiltration for a two layered soil under complex rainfall patterns. *J. Hydrol.* **2000**, *237*, 58–73. [CrossRef]

44. Corradini, C.; Morbidelli, R.; Flammini, A.; Govindaraju, R.S. A parameterized model for local infiltration in two-layered soils with a more permeable upper layer. *J. Hydrol.* **2011**, *396*, 221–232. [CrossRef]

45. Chow, V.T.; Maidment, D.R.; Mays, L.W. *Applied Hydrology*; McGraw-Hill: New York, NY, USA, 1988; ISBN 978-0071001748.

46. Nielsen, D.R.; Biggar, J.W.; Erh, K.T. Spatial variability of field measured soil-water properties. *Hilgardia* **1973**, *42*, 215–259. [CrossRef]

47. Warrick, A.W.; Nielsen, D.R. *Spatial Variability of Soil Physical Properties in the Field. Applications of Soil Physics*; Hillel, D., Ed.; Academic Press: New York, NY, USA, 1980; pp. 319–344.

48. Peck, A.J. Field variability of soil physical properties. In *Advances in Irrigation*; Hillel, D., Ed.; Academic: San Diego, CA, USA, 1983; Volume 2, pp. 189–221, ISBN 0-12-024302-4.

49. Greminger, P.J.; Sud, Y.K.; Nielsen, D.R. Spatial variability of field measured soil-water characteristics. *Soil Sci. Soc. Am. J.* **1985**, *49*, 1075–1082. [CrossRef]

50. Sharma, M.L.; Barron, R.J.W.; Fernie, M.S. Areal distribution of infiltration parameters and some soil physical properties in lateritic catchments. *J. Hydrol.* **1987**, *94*, 109–127. [CrossRef]

51. Loague, K.; Gander, G.A. R-5 revisited, 1. Spatial variability of infiltration on a small rangeland catchment. *Water Resour. Res.* **1990**, *26*, 957–971. [CrossRef]

52. Logsdon, S.D.; Jaynes, D.B. Spatial variability of hydraulic conductivity in a cultivated field at different times. *Soil Sci. Soc. Am. J.* **1996**, *60*, 703–709. [CrossRef]

53. Russo, D.; Bresler, E. Soil hydraulic properties as stochastic processes: 1. Analysis of field spatial variability. *Soil Sci. Soc. Am. J.* **1981**, *45*, 682–687. [CrossRef]

54. Russo, D.; Bresler, E. A univariate versus a multivariate parameter distribution in a stochastic-conceptual analysis of unsaturated flow. *Water Resour. Res.* **1982**, *18*, 483–488. [CrossRef]

55. Sharma, M.L.; Seely, E. Spatial variability and its effect on infiltration. In Proceedings of the Hydrology Water Resources Symposium; Institute of Engineering: Perth, Australia, 1979; pp. 69–73.

56. Maller, R.A.; Sharma, M.L. An analysis of areal infiltration considering spatial variability. *J. Hydrol.* **1981**, *52*, 25–37. [CrossRef]

57. Saghafian, B.; Julien, P.Y.; Ogden, F.L. Similarity in catchment response: 1. Stationary rainstorms. *Water Resour. Res.* **1995**, *31*, 1533–1541. [CrossRef]

58. Sivapalan, M.; Wood, E.F. Spatial heterogeneity and scale in the infiltration response of catchments. In *Scale Problems in Hydrology*; Water Science and Technology Library; Gupta, V.K., Rodríguez-Iturbe, I., Wood, E.F., Eds.; D. Reidel Publishing: Dordrecht, Holland, 1986; pp. 81–106, ISBN 978-94-010-8579-3.

59. McKay, M.D.; Beckman, R.J.; Connover, W.J. A comparison of three methods for selecting values of input variables in the analysis of output from a computer code. *Technometrics* **1979**, *21*, 239–245. [CrossRef]

60. Smith, R.E.; Goodrich, D.C. Model for rainfall excess patterns on randomly heterogeneous area. *J. Hydrol. Eng.* **2000**, *5*, 355–362. [CrossRef]

61. Govindaraju, R.S.; Morbidelli, R.; Corradini, C. Areal infiltration modeling over soils with spatially-correlated hydraulic conductivities. *J. Hydrol. Eng.* **2001**, *6*, 150–158. [CrossRef]

62. Corradini, C.; Govindaraju, R.S.; Morbidelli, R. Simplified modelling of areal average infiltration at the hillslope scale. *Hydrol. Proc.* **2002**, *16*, 1757–1770. [CrossRef]

63. Smith, R.E.; Hebbert, R.H.B. A Monte Carlo analysis of the hydrologic effects of spatial variability of infiltration. *Water Resour. Res.* **1979**, *15*, 419–429. [CrossRef]

64. Woolhiser, D.A.; Smith, R.E.; Giraldez, J.-V. Effects of spatial variability of saturated hydraulic conductivity on Hortonian overland flow. *Water Resour. Res.* **1996**, *32*, 671–678. [CrossRef]

65. Corradini, C.; Morbidelli, R.; Melone, F. On the interaction between infiltration and Hortonian runoff. *J. Hydrol.* **1998**, *204*, 52–67. [CrossRef]

66. Nahar, N.; Govindaraju, R.S.; Corradini, C.; Morbidelli, R. Role of run-on for describing field-scale infiltration and overland flow over spatially variable soils. *J. Hydrol.* **2004**, *286*, 36–51. [CrossRef]

67. Langhans, C.; Govers, G.; Diels, J. Development and parameterization of an infiltration model accounting for water depth and rainfall intensity. *Hydrol. Proc.* **2013**, *27*, 3777–3790. [CrossRef]

68. Goodrich, D.C.; Faurès, J.-M.; Woolhiser, D.A.; Lane, L.J.; Sorooshian, S. Measurement and analysis of small-scale convective storm rainfall variability. *J. Hydrol.* **1995**, *173*, 283–308. [CrossRef]

69. Krajewski, W.F.; Ciach, G.J.; Habib, E. An analysis of small-scale rainfall variability in different climatic regimes. *Hydrol. Sci. J.* **2003**, *48*, 151–162. [CrossRef]

70. Wood, E.F.; Sivapalan, M.; Beven, K. Scale effects in infiltration and runoff production. In *Proceedings of the Symposium on Conjunctive Water Use*; IAHS: Budapest, Hungary, 1986; ISBN 0-947571-85-X.

71. Castelli, F. A simplified stochastic model for infiltration into a heterogeneous soil forced by random precipitation. *Adv. Water Resour.* **1996**, *19*, 133–144. [CrossRef]

72. Govindaraju, R.S.; Corradini, C.; Morbidelli, R. A semi-analytical model of expected areal-average infiltration under spatial heterogeneity of rainfall and soil saturated hydraulic conductivity. *J. Hydrol.* **2006**, *316*, 184–194. [CrossRef]

73. Morbidelli, R.; Corradini, C.; Govindaraju, R.S. A field-scale infiltration model accounting for spatial heterogeneity of rainfall and soil saturated hydraulic conductivity. *Hydrol. Proc.* **2006**, *20*, 1465–1481. [CrossRef]

74. Corradini, C.; Flammini, A.; Morbidelli, R.; Govindaraju, R.S. A conceptual model for infiltration in two-layered soils with a more permeable upper layer: From local to field scale. *J. Hydrol.* **2011**, *410*, 62–72. [CrossRef]

75. Govindaraju, R.S.; Corradini, C.; Morbidelli, R. Local and field-scale infiltration into vertically non-uniform soils with spatially-variable surface hydraulic conductivities. *Hydrol. Proc.* **2012**, *26*, 3293–3301. [CrossRef]

76. Beven, K.J. *Rainfall-Runoff Modelling*; John Wiley & Sons: Hoboken, NJ, USA, 2002; ISBN 978-0-470-71459-1.

77. Fiori, A.; Romanelli, M.; Cavalli, D.J.; Russo, D. Numerical experiments of streamflow generation in steep catchments. *J. Hydrol.* **2007**, *339*, 183–192. [CrossRef]

78. Nassif, S.H.; Wilson, E.M. The influence of slope and rain intensity on runoff and infiltration. *Hydrol. Sci. Bull.* **1975**, *20*, 539–553. [CrossRef]

79. Sharma, K.; Singh, H.; Pareek, O. Rainwater infiltration into a bar loamy sand. *Hydrol. Sci. J.* **1983**, *28*, 417–424. [CrossRef]

80. Poesen, J. The influence of slope angle on infiltration rate and hortonian overland flow volume. *Z. Geomorphol.* **1984**, *49*, 117–131.

81. Philip, J.R. Hillslope infiltration: Planar slopes. *Water Resour. Res.* **1991**, *27*, 109–117. [CrossRef]

82. Cerdà, A.; García-Fayos, P. The influence of slope angle on sediment, water and seed losses on badland landscapes. *Geomorphology* **1997**, *18*, 77–90. [CrossRef]

83. Fox, D.M.; Bryan, R.B.; Price, A.G. The influence of slope angle on final infiltration rate for interrill conditions. *Geoderma* **1997**, *80*, 181–194. [CrossRef]

84. Chaplot, V.; Le Bissonais, Y. Field measurements of interrill erosion under different slopes and plot sizes. *Earth Surf. Process. Landf.* **2000**, *25*, 145–153. [CrossRef]

85. Janeau, J.L.; Bricquet, J.P.; Planchon, O.; Valentin, C. Soil crusting and infiltration on steep slopes in northern Thailand. *Eur. J. Soil Sci.* **2003**, *54*, 543–553. [CrossRef]

86. Assouline, S.; Ben-Hur, M. Effects of rainfall intensity and slope gradient on the dymanics of interrill erosion during soil surface sealing. *Catena* **2006**, *66*, 211–220. [CrossRef]

87. Chen, L.; Young, M.H. Green-Ampt infiltration model for sloping surfaces. *Water Resour. Res.* **2006**, *42*, W07420. [CrossRef]

88. Essig, E.T.; Corradini, C.; Morbidelli, R.; Govindaraju, R.S. Infiltration and deep flow over sloping surfaces: Comparison of numerical and experimental results. *J. Hydrol.* **2009**, *374*, 30–42. [CrossRef]

89. Ribolzi, O.; Patin, J.; Bresson, L.; Latsachack, K.; Mouche, E.; Sengtaheuanghoung, O.; Silvera, N.; Thiébaux, J.P.; Valentin, C. Impact of slope gradient on soil surface features and infiltration on steep slopes in northern Laos. *Geomorphology* **2011**, *127*, 53–63. [CrossRef]

90. Patin, J.; Mouche, E.; Ribolzi, O.; Chaplot, V.; Sengtaheuanghoung, O.; Latsachak, K.O.; Soulileuth, B.; Valentin, C. Analysis of runoff production at the plot scale during a long-term survey of a small agricultural catchment in Lao PDR. *J. Hydrol.* **2012**, *426–427*, 79–92. [CrossRef]

91. Lv, M.; Hao, Z.; Liu, Z.; Yu, Z. Conditions for lateral downslope unsaturated flow and effects of slope angle on soil moisture movement. *J. Hydrol.* **2013**, *486*, 321–333. [CrossRef]

92. Morbidelli, R.; Saltalippi, C.; Flammini, A.; Cifrodelli, M.; Corradini, C.; Govindaraju, R.S. Infiltration on sloping surfaces: Laboratory experimental evidence and implications for infiltration modelling. *J. Hydrol.* **2015**, *523*, 79–85. [CrossRef]

93. Mu, W.; Yu, F.; Li, C.; Xie, Y.; Tian, J.; Liu, J.; Zhao, N. Effects of rainfall intensity and slope gradient on runoff and soil moisture content on different growing stages of spring maize. *Water* **2015**, *7*, 2990–3008. [CrossRef]

94. Khan, M.N.; Gong, Y.; Hu, T.; Lal, R.; Zheng, J.; Justine, M.F.; Azhar, M.; Che, M.; Zhang, H. Effect of slope, rainfall intensity and mulch on erosion and infiltration under simulated rain on purple soil of south-western Sichuan province, China. *Water* **2016**, *8*, 528. [CrossRef]

95. Morbidelli, R.; Saltalippi, C.; Flammini, A.; Cifrodelli, M.; Picciafuoco, T.; Corradini, C.; Govindaraju, R.S. Laboratory investigation on the role of slope on infiltration over grassy soils. *J. Hydrol.* **2016**, *543*, 542–547. [CrossRef]

96. Wang, J.; Chen, L.; Yu, Z. Modeling rainfall infiltration on hillslopes using flux-concentration relation and time compression approximation. *J. Hydrol.* **2018**, *557*, 243–253. [CrossRef]

97. Morbidelli, R.; Saltalippi, C.; Flammini, A.; Govindaraju, R.S. Role of slope on infiltration: A review. *J. Hydrol.* **2018**, *557*, 878–886. [CrossRef]

98. Gerke, H.H. Preferential flow description for structured soils. *J. Plant Nutr. Soil Sci.* **2006**, *169*, 382–400. [CrossRef]

99. Simunek, J.; Jarvis, N.J.; van Genuchten, M.T.; Gardenas, A. Review and comparison of models describing non-equilibrium and preferential flow and transport in the vadose zone. *J. Hydrol.* **2003**, *272*, 14–35. [CrossRef]

100. Buckingham, E. *Studies on the Movement of Soil Moisture*; US Department of Agriculture, Bureau of Soils No. 38; USDA: Washington, DC, USA, 1907.

101. Brooks, R.H.; Corey, A.T. *Hydraulic Properties of Porous Media*; Hydrology Paper; Colorado State University: Fort Collins, CO, USA, 1964; Volume 3, p. 27.

102. Van Genuchten, M.T. A closed-form equation for predicting the hydraulic conductivity of unsaturated soils. *Soil Sci. Soc. Am. J.* **1980**, *44*, 892–898. [CrossRef]

103. Kacimov, A.; Obnosov, Y. Pseudo-hysteretic double-front hiatus-stage soil water parcels supplying a plant-root continuum: The Green-Ampt-Youngs model revisited. *Hydrol. Sci. J.* **2013**, *58*, 237–248. [CrossRef]

104. Mein, R.G.; Larson, C.L. Modeling infiltration during a steady rain. *Water Resour. Res.* **1973**, *9*, 384–394. [CrossRef]

105. Chu, B.T.; Parlange, J.-Y.; Aylor, D.E. Edge effects in linear diffusion. *Acta Mech.* **1978**, *21*, 13–27. [CrossRef]

106. Parlange, J.-Y.; Lisle, I.; Braddock, R.D.; Smith, R.E. The three-parameter infiltration equation. *Soil Sci.* **1982**, *133*, 337–341. [CrossRef]

107. Eagleson, P.S. *Dynamic Hydrology*; McGraw-Hill: New York, NY, USA, 1970.

108. Raudkivi, A.J. *Hydrology*; Pergamon: Oxford, UK, 1979; ISBN 978-0080242613.

109. Philip, J.R. Theory of infiltration. In *Advances in Hydroscience*; Chow, W.T., Ed.; Academic Press: New York, NY, USA, 1969; Volume 5, pp. 215–296.

110. Youngs, E.G. An infiltration method measuring the hydraulic conductivity of unsaturated porous materials. *Soil Sci.* **1964**, *97*, 307–311. [CrossRef]

111. Mls, J. Effective rainfall estimation. *J. Hydrol.* **1980**, *45*, 305–311. [CrossRef]

112. Péschke, G.; Kutilek, M. Infiltration model in simulated hydrographs. *J. Hydrol.* **1982**, *56*, 369–379. [CrossRef]

113. Verma, S.C. Modified Horton's infiltration equation. *J. Hydrol.* **1982**, *58*, 383–388. [CrossRef]

114. Reeves, M.; Miller, E.E. Estimating infiltration for erratic rainfall. *Water Resour. Res.* **1975**, *11*, 102–110. [CrossRef]

115. Basha, H.A. Infiltration models for semi-infinite soil profiles. *Water Resour. Res.* **2011**, *47*, W08516. [CrossRef]

116. Basha, H.A. Infiltration models for soil profiles bounded by a water table. *Water Resour. Res.* **2011**, *47*, W10527. [CrossRef]

117. Melone, F.; Corradini, C.; Morbidelli, R.; Saltalippi, C. Laboratory experimental check of a conceptual model for infiltration under complex rainfall patterns. *Hydrol. Proc.* **2006**, *20*, 439–452. [CrossRef]

118. Melone, F.; Corradini, C.; Morbidelli, R.; Saltalippi, C.; Flammini, A. Comparison of theoretical and experimental soil moisture profiles under complex rainfall patterns. *J. Hydrol. Eng.* **2008**, *13*, 1170–1176. [CrossRef]

119. Mualem, Y.; Assouline, S. Modeling soil seal as a nonuniform layer. *Water Resour. Res.* **1989**, *25*, 2101–2108. [CrossRef]

120. Bullock, M.S.; Temper, W.D.; Nelson, S.D. Soil cohesion as affected by freezing, water content, time and tillage. *Soil Sci. Soc. Am. J.* **1988**, *52*, 770–776. [CrossRef]

121. Emmerich, W.E. Season and erosion pavement influence on saturated soil hydraulic conductivity. *Soil Sci.* **2003**, *168*, 637–645. [CrossRef]

122. Morbidelli, R.; Corradini, C.; Saltalippi, C.; Flammini, A.; Rossi, E. Infiltration-soil moisture redistribution under natural conditions: Experimental evidence as a guideline for realizing simulation models. *Hydrol. Earth Syst. Sci.* **2011**, *15*, 2937–2945. [CrossRef]

123. Chen, L.; Sela, S.; Svoray, T.; Assouline, S. The role of soil-surface sealing, microtopography, and vegetation patches in rainfall-runoff processes in semiarid areas. *Water Resour. Res.* **2013**, *49*, 5585–5599. [CrossRef]

124. Morbidelli, R.; Saltalippi, C.; Flammini, A.; Rossi, E.; Corradini, C. Soil water content vertical profiles under natural conditions: Matching of experiments and simulations by a conceptual model. *Hydrol. Proc.* **2014**, *28*, 4732–4742. [CrossRef]

125. Singh, V.P. *Kinematic Wave Modeling in Water Resources: Surface Water Hydrology*; John Wiley & Sons: New York, NY, USA, 1996; ISBN 0-471-10945-2.

126. Morbidelli, R.; Corradini, C.; Saltalippi, C.; Brocca, L. Initial soil water content as input to field-scale infiltration and surface runoff models. *Water Resour. Manag.* **2012**, *26*, 1793–1807. [CrossRef]

127. Maidment, D.R. *Handbook of Hydrology*; McGraw-Hill: New York, NY, USA, 1993; ISBN 9780070397323.

128. Bronstert, A.; Bardossy, A. The role of spatial variability of soil moisture for modelling surface runoff generation at the small catchment scale. *Hydrol. Earth Syst. Sci.* **1999**, *3*, 505–516. [CrossRef]

129. Beven, K.; Germann, P. Macropores and water flow in soils revisited. *Water Resour. Res.* **2013**, *49*, 3071–3092. [CrossRef]

130. Simunek, J.; van Genuchten, M.T. Modeling nonequilibrium flow and transport processes using HYDRUS. *Vadose Zone J.* **2008**, *7*, 782–797. [CrossRef]

131. Ireson, A.M.; Butler, A.P. Controls on preferential recharge to Chalk aquifers. *J. Hydrol.* **2011**, *398*, 109–123. [CrossRef]

132. Van Dam, J.C.; Groenendijk, P.; Hendriks, R.F.A.; Kroes, J.G. Advances of modeling water flow in variability saturated soils with SWAP. *Vadose Zone J.* **2008**, *7*, 640–653. [CrossRef]

133. Morbidelli, R.; Saltalippi, C.; Flammini, A.; Cifrodelli, M.; Picciafuoco, T.; Corradini, C.; Govindaraju, R.S. In-situ measurements of soil saturated hydraulic conductivity: Assessment of reliability through rainfall-runoff experiments. *Hydrol. Proc.* **2017**, *31*, 3084–3094. [CrossRef]

134. Flammini, A.; Morbidelli, R.; Saltalippi, C.M.; Picciafuoco, T.; Corradini, C.; Govindaraju, R.S. Reassessment of a semi-analytical field-scale infiltration model through experiments under natural rainfall events. *J. Hydrol.* **2018**, *565*, 835–845. [CrossRef]

Article

A New Conceptual Model for Slope-Infiltration

Renato Morbidelli [1,*], Corrado Corradini [1], Carla Saltalippi [1], Alessia Flammini [1], Jacopo Dari [1] and Rao S. Govindaraju [2]

[1]   Department of Civil and Environmental Engineering, University of Perugia, via G. Duranti 93, 06125 Perugia, Italy; corrado.corradini@unipg.it (C.C.); carla.saltalippi@unipg.it (C.S.); alessia.flammini@unipg.it (A.F.); jacopo.dari@unifi.it (J.D.)

[2]   Lyles School of Civil Engineering, Purdue University, West Lafayette, IN 47907, USA; govind@purdue.edu

*   Correspondence: renato.morbidelli@unipg.it; Tel.: +39-075-5853620

Received: 21 February 2019; Accepted: 28 March 2019; Published: 1 April 2019

Abstract: Rainfall infiltration modeling over surfaces with significant slopes is an unsolved problem. Even though water infiltration occurs over soil surfaces with noticeable gradients in most real situations, the typical mathematical models used were developed for infiltration over horizontal surfaces. In addition, recent investigations on infiltration over sloping surfaces have provided conflicting results, suggesting that our understanding of the process may still be lacking. In this study, our objective is to specifically examine if the surface water velocity that is negligible over near horizontal soil surfaces can affect the infiltration process over steep slopes. A new conceptual model representing a wide range of experimental results is proposed. The model represents water flow as an ensemble of infinitesimal "particles" characterized by specific velocities and assumes that only "particles" with velocity less than a threshold value can contribute to the infiltration process. The velocity distribution and the threshold value depend on slope and soil type, respectively. This conceptual model explains observed results and serves as a foundation for developing further experiments and refining models that offer more realistic representations of infiltration over sloping surfaces.

Keywords: hillslope hydrology; sloping surfaces; infiltration process; infiltration modeling; overland flow

---

1. Introduction

The process of infiltration of water into the soil is highly dependent on soil hydraulic properties that are generally variable in space, both in the vertical and horizontal directions. In natural conditions, the net rainfall reaching the soil is also affected by the vegetation cover that produces rainfall interception, sheltering the soil surface from the impact of falling drops. Vegetation also provides root systems that generate preferential subsurface flow paths.

Historically, solutions to infiltration problems have been represented through analytical, numerical, conceptual and empirical mathematical formulations. Analytical solutions provide estimates of infiltration rate or cumulative infiltration as functions of time, usually by simplifications on the soil water content profile during the study period. Powerful computers use numerical simulations of unsaturated soil domains in a single vertical direction or in multiple spatial dimensions, allowing for the use of complex initial and boundary conditions. Conceptual models try to balance the reduction of process complexity with a satisfactory representation of physical reality, obtaining simplified problem formulations. Finally, empirical infiltration models involve parameters fitted to the measured infiltration, but they have limited power as predictive tools because the same model cannot be used in different catchments.

The infiltration process has been deeply analyzed since early parts of last century at the point (or local) scale and lately also at a field scale, even though most models assumed a horizontal soil surface. At the local scale, when single storms are considered, classical equations (e.g., References [1–6]) or more recent formulations (e.g., Reference [7]) are generally adopted, while in the presence of events with consecutive soil water infiltration-redistribution cycles, conceptual/semi-analytical models, such as the one described by Reference [8], become necessary. Upscaling of these local models to obtain real (or field) models has been performed considering both vertically homogeneous [9–11] and layered [12] soils.

All the aforementioned models, as well as many others in the scientific literature, consider a soil surface that is oriented horizontally, while in most practical conditions the infiltration process occurs on surfaces with significant gradients [13]. The results obtained in the latter condition are not conclusive [14] and a physically-based approach with the ability to justify the experimental results needs to be developed. Table 1 shows a summary of theoretical (both analytical and conceptual) and experimental (carried out in both laboratory and field) studies dealing with the role of slope in infiltration. A comparison of these analyses, especially when carried out in natural fields, is confounded by several factors such as soil type and microtopography, rainfall intensity and duration, and presence of vegetation, to name a few. However, there exist laboratory experiments (e.g., Reference [15]) designed to exclude the abovementioned effects, showing a significant reduction of infiltration with slope, beyond the value expected when a steady state condition of soil saturation is assumed. This result is more pronounced for bare and clay soils rather than in vegetated and coarse-textured soils. Instead, most studies in Table 1 showing that infiltration increases with increasing slope were characterized by the formation of rills and/or a sealing layer. Nevertheless, some studies reported in Table 1 recommended the adoption of empirical corrections for the saturated hydraulic conductivity, K_s, but these corrections are not based on a theoretical approach and cannot be extended for general use.

Table 1. Studies dealing with the relation between slope and infiltration.

Authors	Paper	Analysis Type
Infiltration increase with increasing slope		
Poesen	[16]	Experim.
Janeau, Bricquet, Planchon, Valentin	[17]	Experim.
Assouline, Ben-Hur	[18]	Experim.
Chen, Young	[19]	Theor.
Ribolzi, Patin, Bresson, Latsachack, Mouche, Sengtaheuanghoung, Silvera, …	[20]	Experim
Infiltration decrease with increasing slope		
Nassif, Wilson	[21]	Experim.
Sharma, Barron, Fernie	[22]	Experim.
Philip	[23]	Theor.
Fox, Bryan, Price	[24]	Experim.
Chaplot, Le Bissonnais	[25]	Experim.
Essig, Corradini, Morbidelli, Govindaraju	[26]	Experim.
Patin, Mouche, Ribolzi, Chaplot, Sengtaheuanghoung, Latsachack, …	[27]	Experim.
Morbidelli, Saltalippi, Flammini, Cifrodelli, Corradini, Govindaraju	[15]	Experim.
Mu, Yu, Li, Xie, Tian, Liu, Zhao	[28]	Experim.
Khan, Gong, Hu, Lai, Zheng, Justine, Azhar, Che, Zhang	[29]	Experim.
Morbidelli, Saltalippi, Flammini, Cifrodelli, Picciafuoco, Corradini, …	[30]	Experim.
Wang, Chen, Yu	[31]	Theor.

Further theoretical developments are needed to understand the complex processes of infiltration over sloping surfaces and to obtain a model that is sufficiently representative of the available experimental results. A clear physical interpretation is required to support the prevailing hypothesis of decreasing infiltration with increasing slope and to indicate specific laboratory experiments needed to assess the proposed model.

The main objective of this paper is to present a conceptual model able to justify the reduction of infiltration (with increasing gradients) obtained in the absence of secondary disturbance effects.

The basic idea of the proposed model would serve as a potential starting point to stimulate the development of new experiments to identify a model closer to physical reality.

2. Basic Equations

A study of the slope-infiltration interaction should be carried out by considering the coupling of the fundamental processes occurring at and immediately below the soil surface (Figure 1), i.e., infiltration and surface runoff.

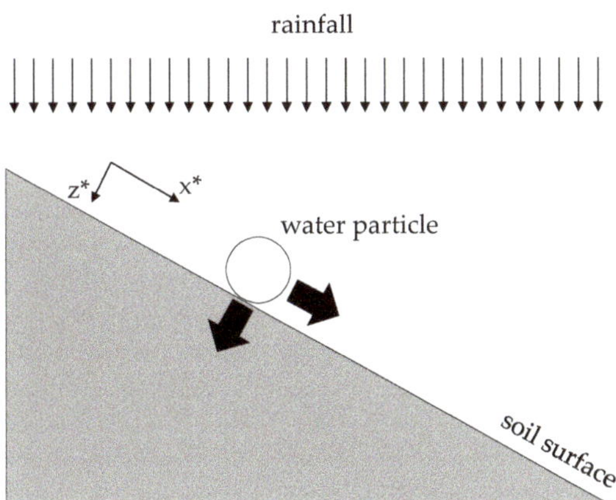

Figure 1. Schematic representation of the infiltration process over a sloping surface in a Cartesian coordinate system.

Under conditions of a homogeneous and isotropic soil and considering a smooth soil surface, the best theoretical representation of the involved processes can be obtained through the Richards equation for the sub-surface flow and the Saint-Venant equations for the overland flow. More specifically, the uni-dimensional Richards equation for a flat surface can be rewritten, in Cartesian coordinates (x^*, z^*) shown in Figure 1, in the form (Reference [23]):

$$\frac{\partial \theta}{\partial t} = \frac{\partial}{\partial z^*}\left[D\frac{\partial \theta}{\partial z^*}\right] - \frac{dK}{d\theta}\frac{\partial \theta}{\partial z^*}\cos \gamma, \tag{1}$$

where t is the time, θ the volumetric soil water content, K the hydraulic conductivity, D the soil water diffusivity equal to $K(\theta)(d\psi(\theta)/d\theta)$ with ψ soil suction head, and γ the slope angle. Equation (1), under appropriate initial and boundary conditions, is identical to the classical Richards equation but with K in the new coordinate system substituted by $K\cos\gamma$.

The Saint-Venant equations, with a lateral inflow q_l per unit length that does not contribute any additional momentum to the flow, may be written as:

$$\frac{\partial Q}{\partial x^*} + \frac{\partial A}{\partial t} = q_l, \tag{2}$$

$$S_f = S_0 - \frac{\partial h}{\partial x^*} - \frac{v}{g}\frac{\partial v}{\partial x^*} - \frac{1}{g}\frac{\partial v}{\partial t}, \tag{3}$$

where Q is the discharge, A the area of a cross-section, v the velocity, S_f the friction slope, S_0 the surface slope, h the flow depth with reference to the bottom, and g the gravity acceleration.

A straightforward analysis of Equations (1)–(3) along with initial and boundary conditions highlights that they cannot interact adequately. Specifically, the solution of the Richards equation may produce effects on the Saint-Venant equations (e.g., when the soil surface is saturated, the rainfall excess determined through Equation (1) and its boundary conditions could be identified with q_l, that moves along the slope according to Equations (2) and (3)). On the other hand, there is no possibility for the Saint-Venant equations to produce, from a mathematical point of view, any effect on the upper boundary conditions of the Richards equation, therefore precluding a possible interaction between the overland flow velocity and infiltration. Currently, we do not have a good way of defining the interface conditions between surface and subsurface flows at the soil surface.

3. The Conceptual Model

The basic element of the proposed model is inspired by the following observation. When a golf player is completing a hole, the last shot over the green has to be realized with a threshold velocity. If the ball is characterized by a velocity greater than this threshold value, it does not enter the hole (Figure 2) even though the direction is correct. A similar effect may be applicable to water moving on a porous surface.

Let us consider steady conditions with a small layer of overland flow generated by a rainfall rate, r. In line with the aforementioned abstraction, water "particles" can be roughly considered like balls running over a soil surface and drawn in the soil pores by gravity but in the presence of an interaction among the liquid water "particles". Furthermore, it is logical to consider that the arrival of new "particles" close to the pores depends on the persistence of overland flow under a given rainfall rate. The last condition assures the existence of "particles" that potentially may fall in the pores.

For a horizontal soil surface, considering that (1) all the "particles" may potentially enter the pores, because their velocity over the surface is practically equal to zero, and (2) the rainfall excess cannot enter the pores, the proposed model becomes unnecessary.

The interest in this conceptual model arises in case the soil surface possesses a slope, as in many natural conditions. With rainfall excess, water particles move downstream. Depending on friction slope, fluid viscosity, and other local conditions, each infinitesimal "particle" of water is characterized by a specific velocity (Figure 3). Consequently, it can be assumed that the "particles" move over the slope with different velocities that in the average increase with the distance from the soil surface.

The quantity "particle velocity" in the small layer that produces subsurface flow may be assumed as a stochastic variable characterized by a specific cumulative probability that for simplicity can be expressed through an exponential term. The last choice, somewhat arbitrary, could be changed without significant alterations of the proposed conceptual model. We have:

$$P(v \leq v_l) = 1 - e^{-\lambda v_l} \tag{4}$$

where λ (>0) is the parameter of the probability distribution, v the independent variable, and v_l a specific value of v representing the maximum velocity that allows infiltration of "particles" in a given pore.

Therefore "particles" with velocity less than v_l may enter a specific pore, while "particles" with a velocity greater than v_l continue their run over the surface. Only the fraction $P(v \leq v_l)$ may fall into the pore producing subsurface flow.

As for a golf player, the threshold velocity depends on the hole diameter. In the proposed conceptual model, v_l depends on the pore dimensions that are linked with soil texture and particle layout in the soil matrix. This means that v_l may be linked with K_s. Furthermore, $P(v \leq v_l)$ changes with increasing slope because of the increasing velocity of "particles". Consequently, for steeper slopes, the probability to have values of v lower than v_l decreases and Equation (4) suggests that λ becomes smaller (Figure 4). The probability $P(v \leq v_l)$ is also influenced, particularly in the presence of vegetation, by the surface roughness, but bare soils are considered here.

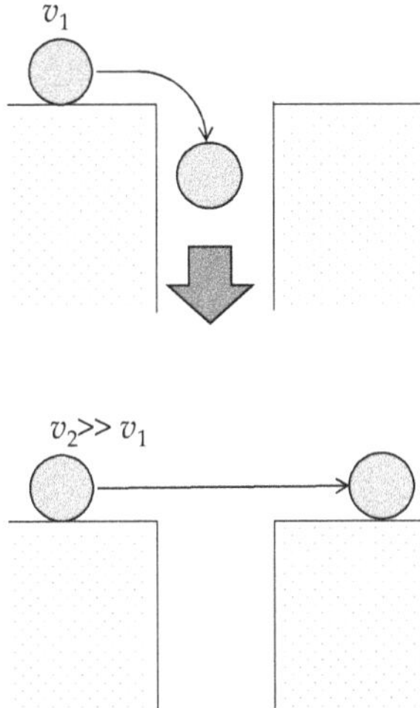

Figure 2. Movement of balls in a golf game: Balls can fall into the hole or not depending on their velocity ($v_2 >> v_1$).

Hence, a given surface slope determines a specific $P(v \leq v_l)$ through a specific λ value, while the soil structure, characterized by a well-known K_s, affects the threshold velocity v_l. As a final result, under steady conditions, infiltration of water into a slope, K_{se}, can be obtained as $K_{se} = K_s \times P(v < v_l)$, with K_s that represents the saturated hydraulic conductivity for a horizontal soil surface. The quantity K_{se} may be considered as an effective saturated hydraulic conductivity depending on the soil gradient.

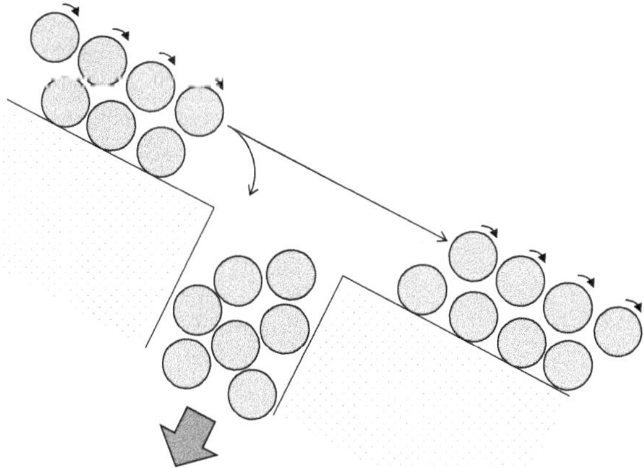

Figure 3. Soil surface with significant slope and "particles" characterized by various velocities.

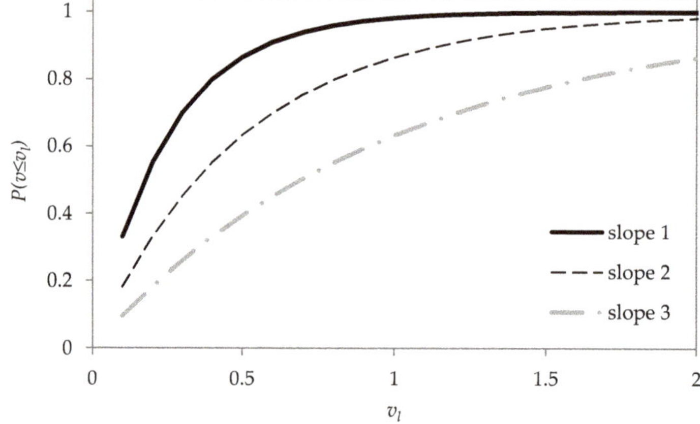

Figure 4. Example of three different cumulative probability functions of the "particle" velocities associated to different slopes (slope 1 < slope 2 < slope 3) with the corresponding λ values in Equation (4) equal to 4, 2, and 1 s/cm, respectively.

4. Model Parameters

The key parameters of the model are the threshold velocity, v_l, and the decay parameter of the cumulative probability, λ. A simple approach to obtain their values can be to fix one of them and to determine the other using a calibration procedure based on the use of experimental data. As discussed above, the threshold velocity is physically linked with the pore diameters and therefore with soil texture and particle layout, which in turn influence K_s. We assume v_l equal to K_s even though this choice will affect the estimate of λ, that should depend only on the slope while its value obtained through calibration will also adjust for this non-optimal hypothesis. Really, this assumption does not reflect the physical reality because v_l depends on the pore diameters that in a given soil is a random variable and therefore should be represented by a stochastic approach. Therefore, our rough simplification with a sole value of v_l equal to K_s is equivalent to considering all soil pores characterized by a representative diameter.

5. Experimental System

The calibration process of the λ parameter requires the use of results obtained by laboratory experiments carried out through a physical model [26]. The adopted equipment is 1.52 m long, 1.22 m wide, and 0.78 m deep, with a tray angle adjustable up to 30°. As it can be seen in Figure 5, all boundaries of the physical model are impermeable except for the soil surface. A small gravel layer (with thickness 7 cm) is placed at the bottom of the soil. Surface and percolated/deep flows are measured by two calibrated sensors based on a tipping-bucket mechanism. Different steady rainfall rates, sufficiently uniform over the slope with values up to 80 mm h^{-1}, can be generated using special sprinklers, a pump, and a manual manometer. The various natural soils used for the experiments are characterized by the grain size distributions shown in Figure 6 with saturated hydraulic conductivity associated to horizontal surface, K_s, equal to 2.93 mm h^{-1}, 3.20 mm h^{-1}, 10.37 mm h^{-1} and 17.00 mm h^{-1} for soil 1, soil 2, soil 3, and soil 4, respectively [26,32–36]. Each experimental run lasted more than 24 h, and the rainfall that was applied during the first 8–10 h produced extended periods with steady conditions in the absence of direct rainfall infiltration. Before the beginning of each experiment, rainfall was applied in order to have high soil water content at any depth. Furthermore, in the time period between two successive experiments, the surface water content was kept sufficiently high in order to avoid the formation of cracks. The measurements of deep flow for different gradients obtained under steady conditions are assumed equal to K_{se}. Soil erosion did not affect the experiments as checked by an analysis of the surface water cloudiness.

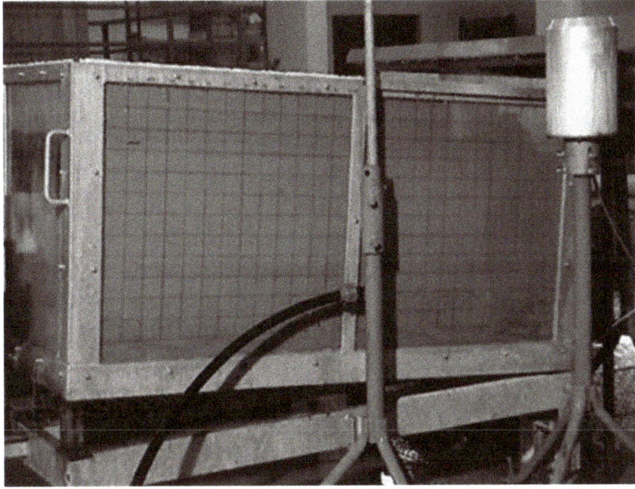

Figure 5. A view of the physical model adopted in the laboratory experiments for slope-infiltration studies.

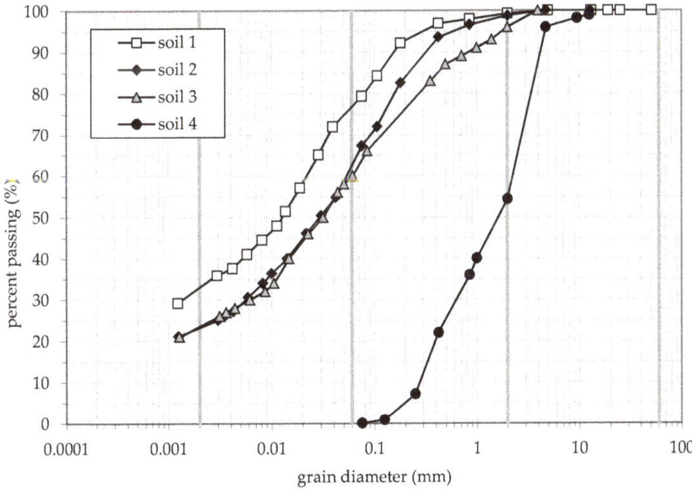

Figure 6. Grain size distribution of the soils used in the laboratory experiments.

6. Analysis of Results

Some experimental results on the laboratory system described above were deduced by Reference [26]. Three natural soils were used in different sets of laboratory experiments. The first set of experiments (performed by using soil 1 of Figure 6, labeled as "Clay Loam" soil according to USDA soil classification) consisted of 24 trials with slopes in the range 1°–15° and rainfall rates in the range 10–20 mm h^{-1}. The second set (by using soil 2 of Figure 6, labeled as "Loam" soil by USDA) consisted of 8 trials with slopes and rainfall rates in the ranges 1°–15° and 10–15 mm h^{-1}, respectively. Finally, the third set of experiments (soil 3 of Figure 6, labeled as "Sandy Loam" soil by USDA) consisted of 18 trials where slopes and rainfall rates were in the ranges 1°–10° and 20–30 mm h^{-1}, respectively. As it can be seen, the adopted surface slopes and soil types were not exhaustive because they didn't involve slopes exceeding 15° and didn't consider a very coarse textured soil.

Therefore, a new set of experiments has been carried out with a fourth soil type (soil 4 of Figure 6, labeled as "Sandy" soil by USDA) considering the surface slope in the range 10°–26° and a very heavy rainfall rate (about 60 mm h^{-1}). The main results obtained by these new experiments are summarized in Table 2 and confirm that for a fixed reference rainfall rate the slope gradient has a negative influence on the steady deep flow. These results are consistent with those of Reference [26]. Even if they refer to conditions dominated by gravitational effects, the theoretical formulations earlier proposed provide incorrect simulations.

Table 2. Steady surface and deep flows for different laboratory slope gradients under a rainfall rate of about 60 mm h^{-1}. Soil 4 of Figure 6 is shown.

Experiment Number	Slope (°)	Average Rainfall Rate (mm h^{-1})	Steady Surface Flow		Steady Deep Flow	
			(mm h^{-1})	(%)	(mm h^{-1})	(%)
1	10	59.10	46.30	78.34	12.80	21.66
2	17	59.36	48.36	81.46	11.00	18.54
3	21	61.10	52.60	86.09	8.50	13.91
4	26	62.60	55.40	88.50	7.20	11.50

The estimate of λ requires us to subdivide the experiments selected here, excluding those carried out with almost horizontal soil surface that have been used for the determination of K_s (14 experiments with slope equal to 1°, as specified in References [15] and [26]), into two groups (with comparable general features) to be used in the calibration and validation phases, respectively (Table 3).

Through an inverse procedure, for each laboratory experiment of the selected calibration set a specific λ parameter can be derived from Equation (4). Specifically, as a first approximation, v_l is assumed to be equal to K_s because a better approach closer to physical reality would require a joint solution of Equations (1) and (2), which is blocked, as pointed out in Section 2, by the difficult representation of their interaction. Then, the term on the left, $P(v \leq v_l)$, is expressed by the ratio between the observed steady deep flow, d_f, and K_s. Therefore, λ is the only unknown quantity in Equation (4).

Considering the laboratory experiments 1–20 of Table 3, the following general relation between slope and λ has been obtained:

$$\lambda = 0.9861e^{-0.139 \times slope}, \tag{5}$$

where the *slope* is expressed in (°). Figure 7 shows the accuracy level of this interpolating function. The scattering of the λ values at given slope angles may be linked with other influential factors not represented in our approach such as, for example, the surface roughness that is expected to be variable from bare soil to another because of their different structure.

The model validation has been made using the laboratory experiments 21–40 of Table 3, through the relative error of the steady deep flow, ε_{df}, defined as follows:

$$\varepsilon_{df} = 100 \frac{\left(K_{se} - d_f\right)}{d_f}, \tag{6}$$

with K_{se} given by:

$$K_{se} = K_s\left(1 - e^{-\lambda K_s}\right). \tag{7}$$

As shown in Table 4, ε_{df} is within the range of −28.9%/+22.6% with an algebraic mean value of 1.0% (or 9.9% when the absolute value of each ε_{df} is considered). In a few cases, the relative error is significant, and therefore, we note that the steady deep flow is generally well reproduced by the proposed model (see also Figure 8).

Finally, from the experimental results, it comes out that increasing the rainfall rate becomes larger for the steady surface flow while the steady deep flow experiences minor changes. More specifically, combining for each slope the results shown in Table 4 for the steady surface and deep flow with the values of rainfall rate given in Table 3, it can be deduced as doubled values of rainfall rate and surface runoff do not determine in the average a clear trend of the deep flow. This outcome may be ascribed to the fact that increasing the water depth on the surface increases its average velocity on the slope, but the velocity of the small layer that affects infiltration does not experience significant variation.

Table 3. Main characteristics of the selected laboratory experiments (from Reference [26] and Table 2), subdivided into calibration and validation sets.

Order Number	Experiment Identification	Slope (°)	Average Rainfall (mm h^{-1})
	Calibration experiments		
1	soil 1, exp 3 [26]	5	9.74
2	soil 1, exp 4 [26]	5	15.16
3	soil 1, exp 19 [26]	5	20.49
4	soil 1, exp 5 [26]	10	10.07
5	soil 1, exp 6 [26]	10	14.62
6	soil 1, exp 21 [26]	10	20.04
7	soil 1, exp 11 [26]	15	8.85
8	soil 1, exp 12 [26]	15	13.18
9	soil 1, exp 23 [26]	15	20.49
10	soil 2, exp 3 [26]	5	9.77
11	soil 2, exp 5 [26]	10	9.91
12	soil 2, exp 7 [26]	15	9.96
13	soil 3, exp 15 [26]	5	18.82
14	soil 3, exp 5 [26]	5	25.86
15	soil 3, exp 7 [26]	5	32.12
16	soil 3, exp 17 [26]	10	18.61
17	soil 3, exp 9 [26]	10	25.82
18	soil 3, exp 11 [26]	10	30.37
19	soil 4, exp 3 (Table 2)	21	61.10
20	soil 4, exp 4 (Table 2)	26	62.60
	Validation experiments		
21	soil 1, exp 13 [26]	5	9.55
22	soil 1, exp 14 [26]	5	13.81
23	soil 1, exp 20 [26]	5	20.28
24	soil 1, exp 9 [26]	10	9.11
25	soil 1, exp 10 [26]	10	14.56
26	soil 1, exp 22 [26]	10	20.02
27	soil 1, exp 7 [26]	15	9.85
28	soil 1, exp 8 [26]	15	14.30
29	soil 1, exp 24 [26]	15	19.84
30	soil 2, exp 4 [26]	5	15.92
31	soil 2, exp 6 [26]	10	15.02
32	soil 2, exp 8 [26]	15	13.90
33	soil 3, exp 16 [26]	5	18.29
34	soil 3, exp 6 [26]	5	25.83
35	soil 3, exp 8 [26]	5	32.31
36	soil 3, exp 18 [26]	10	18.36
37	soil 3, exp 10 [26]	10	25.87
38	soil 3, exp 12 [26]	10	31.34
39	soil 4, exp 1 (Table 2)	10	59.10
40	soil 4, exp 2 (Table 2)	17	59.36

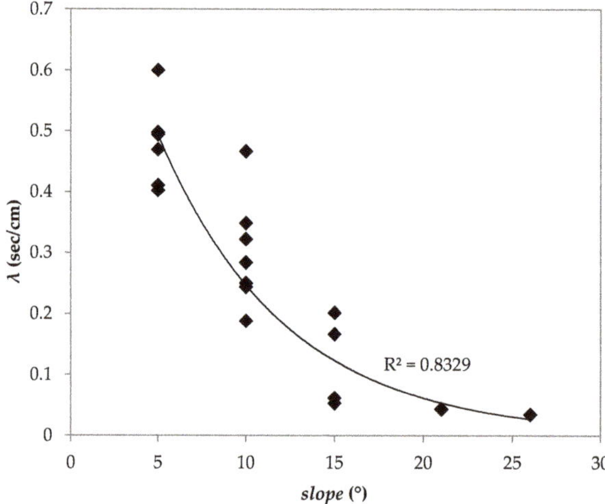

Figure 7. The λ parameter of Equation (4) obtained for the calibration experiments performed at the different slope angles. The best interpolating function is also plotted.

Table 4. Model validation synthesized through the relative error of the steady deep flow, ε_{df} (Equations (6) and (7)).

Order Number	Experiment Identification	Slope (°)	Steady Deep Flow (mm h^{-1})	$K_s \times P(v < v_l)$ (mm h^{-1})	ε_{df} (%)
21	soil 1, exp 13 [26]	5	2.46	2.24	−9.1
22	soil 1, exp 14 [26]	5	2.32	2.24	−3.6
23	soil 1, exp 20 [26]	5	2.11	2.24	6.0
24	soil 1, exp 9 [26]	10	1.38	1.50	8.9
25	soil 1, exp 10 [26]	10	1.24	1.50	21.2
26	soil 1, exp 22 [26]	10	1.79	1.50	−16.0
27	soil 1, exp 7 [26]	15	0.91	0.88	−2.9
28	soil 1, exp 8 [26]	15	0.77	0.88	14.8
29	soil 1, exp 24 [26]	15	0.77	0.88	14.8
30	soil 2, exp 4 [26]	5	2.54	2.54	−0.1
31	soil 2, exp 6 [26]	10	2.27	1.74	−23.3
32	soil 2, exp 8 [26]	15	1.46	1.04	−28.9
33	soil 3, exp 16 [26]	5	9.63	10.31	7.0
34	soil 3, exp 6 [26]	5	10.33	10.31	−0.2
35	soil 3, exp 8 [26]	5	10.25	10.31	0.6
36	soil 3, exp 18 [26]	10	9.46	9.56	1.0
37	soil 3, exp 10 [26]	10	9.54	9.56	0.2
38	soil 3, exp 12 [26]	10	10.05	9.56	−4.9
39	soil 4, exp 1 (Table 2)	10	12.80	15.68	22.6
40	soil 4, exp 2 (Table 2)	17	11.00	12.38	12.5

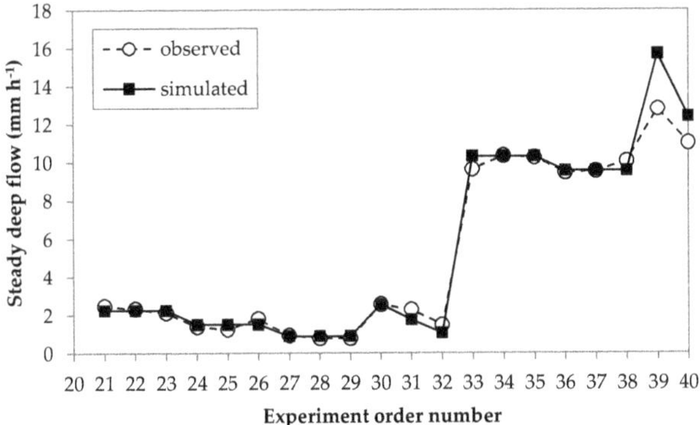

Figure 8. Comparison of the observed and simulated steady deep flow for the different validation experiments.

7. Conclusions

The starting point of this work is that the infiltration process on sloping surfaces observed in many laboratory experiments is not correctly represented by the existing theoretical modeling efforts.

On the basis of the main outcomes of this analysis performed using 54 laboratory experiments and a conceptual model, the following insights can be derived:

- The proposed conceptual model can represent most laboratory experimental results obtained for bare soils in the absence of secondary disturbance effects due to erosion and formation of a sealing layer.
- According to this model, the "particles" of water moving above a sloping surface are characterized by velocities considered as realizations of a stochastic variable with a cumulative probability function expressed through an exponential term.
- Only water "particles" with velocity below a threshold value may contribute to the infiltration process.
- In practical terms, from the knowledge of surface slope, as well as of soil texture and particles layout (both linked to K_s), a cumulative probability function $P(v < v_l)$ may be derived. The behavior of different soil types is represented through the associated K_s values.
- Under steady state conditions, the infiltrating water is given by an effective saturated hydraulic conductivity expressed by $K_s \times P(v < v_l)$.
- Processes related to the fluid viscosity are not explicitly considered.
- The surface roughness plays an important role on the surface flow velocity. When particular vegetation is present (e.g., grassy soil) a specific calibration of λ, which is the unique parameter of the proposed model, is necessary. On this basis to generalize the model, new laboratory experiments with different surface types should be performed.
- This work supports the idea that the infiltration process on a sloping surface is highly conditioned by the surface flow velocity. When the surface slope increases the water speed increases, while infiltration decreases. An extension of this simple concept could be useful to further our knowledge of the infiltration process over sloping surfaces even under unsteady conditions when K_{se} could be used to substitute K_s.

Author Contributions: Laboratory experiments, numerical analyses, investigation, writing—original draft preparation and writing—review and editing, R.M., C.C., C.S., A.F., J.D. and R.S.G.

Funding: This research was partly financed by the Italian Ministry of Education, University and Research (PRIN 2015).

Acknowledgments: The authors are thankful to A.G. Del Prete for his technical assistance.

Conflicts of Interest: The authors declare no conflict of interest.

References

1. Green, W.A.; Ampt, G.A. Studies on soil physics: 1. The flow of air and water through soils. *J. Agric. Sci.* **1911**, *4*, 1–24. [CrossRef]
2. Horton, R.E. An approach toward a physical interpretation of infiltration-capacity. *Soil Sci. Soc. Am. J.* **1940**, *5*, 399–417. [CrossRef]
3. Philip, J.R. The theory of infiltration: 1. The infiltration equation and its solution. *Soil Sci.* **1957**, *83*, 345–357. [CrossRef]
4. Philip, J.R. The theory of infiltration: 2. The profile at infinity. *Soil Sci.* **1957**, *83*, 435–448. [CrossRef]
5. Philip, J.R. The theory of infiltration: 4. Sorptivity algebraic infiltration equation. *Soil Sci.* **1957**, *84*, 257–264. [CrossRef]
6. Smith, R.E.; Parlange, J.-Y. A parameter-efficient hydrologic infiltration model. *Water Resour. Res.* **1978**, *14*, 533–538. [CrossRef]
7. Corradini, C.; Morbidelli, R.; Flammini, A.; Govindaraju, R.S. A parameterized model for local infiltration in two-layered soils with a more permeable upper layer. *J. Hydrol.* **2011**, *396*, 221–232. [CrossRef]
8. Corradini, C.; Melone, F.; Smith, R.E. A unified model for infiltration and redistribution during complex rainfall patterns. *J. Hydrol.* **1997**, *192*, 104–124. [CrossRef]
9. Smith, R.E.; Goodrich, D.C. Model for rainfall excess patterns on randomly heterogeneous area. *J. Hydrol. Eng.* **2000**, *5*, 355–362. [CrossRef]
10. Govindaraju, R.S.; Morbidelli, R.; Corradini, C. Areal infiltration modeling over soils with spatially-correlated hydraulic conductivities. *J. Hydrol. Eng.* **2001**, *6*, 150–158. [CrossRef]
11. Corradini, C.; Govindaraju, R.S.; Morbidelli, R. Simplified modelling of areal average infiltration at the hillslope scale. *Hydrol. Proc.* **2002**, *16*, 1757–1770. [CrossRef]
12. Corradini, C.; Flammini, A.; Morbidelli, R.; Govindaraju, R.S. A conceptual model for infiltration in two-layered soils with a more permeable upper layer: From local to field scale. *J. Hydrol.* **2011**, *410*, 62–72. [CrossRef]
13. Beven, K.J. *Rainfall-Runoff Modelling*; John Wiley & Sons: Chichester, UK, 2002; ISBN 978-0-470-71459-1.
14. Morbidelli, R.; Saltalippi, C.; Flammini, A.; Govindaraju, R.S. Role of slope on infiltration: A review. *J. Hydrol.* **2018**, *557*, 878–886. [CrossRef]
15. Morbidelli, R.; Saltalippi, C.; Flammini, A.; Cifrodelli, M.; Corradini, C.; Govindaraju, R.S. Infiltration on sloping surfaces: Laboratory experimental evidence and implications for infiltration modelling. *J. Hydrol.* **2015**, *523*, 79–85. [CrossRef]
16. Poesen, J. The influence of slope angle on infiltration rate and hortonian overland flow volume. *Z. Geomorphol.* **1984**, *49*, 117–131.
17. Janeau, J.L.; Bricquet, J.P.; Planchon, O.; Valentin, C. Soil crusting and infiltration on steep slopes in northern Thailand. *Europ. J. Soil Sci.* **2003**, *54*, 543–553. [CrossRef]
18. Assouline, S.; Ben-Hur, M. Effects of rainfall intensity and slope gradient on the dymanics of interrill erosion during soil surface sealing. *Catena* **2006**, *66*, 211–220. [CrossRef]
19. Chen, L.; Young, M.H. Green-Ampt infiltration model for sloping surfaces. *Water Resour. Res.* **2006**, *42*, W07420. [CrossRef]
20. Ribolzi, O.; Patin, J.; Bresson, L.; Latsachack, K.; Mouche, E.; Sengtaheuanghoung, O.; Silvera, N.; Thiébaux, J.P.; Valentin, C. Impact of slope gradient on soil surface features and infiltration on steep slopes in northern Laos. *Geomorphology* **2011**, *127*, 53–63. [CrossRef]
21. Nassif, S.H.; Wilson, E.M. The influence of slope and rain intensity on runoff and infiltration. *Hydrol. Sci. Bull.* **1975**, *20*, 539–553. [CrossRef]
22. Sharma, K.; Singh, H.; Pareek, O. Rainwater infiltration into a bar loamy sand. *Hydrol. Sci. J.* **1983**, *28*, 417–424. [CrossRef]
23. Philip, J.R. Hillslope infiltration: Planar slopes. *Water Resour. Res.* **1991**, *27*, 109–117. [CrossRef]
24. Fox, D.M.; Bryan, R.B.; Price, A.G. The influence of slope angle on final infiltration rate for interrill conditions. *Geoderma* **1997**, *80*, 181–194. [CrossRef]

25. Chaplot, V.; Le Bissonais, Y. Field measurements of interrill erosion under different slopes and plot sizes. *Earth Surf. Process. Landf.* **2000**, *25*, 145–153. [CrossRef]
26. Essig, E.T.; Corradini, C.; Morbidelli, R.; Govindaraju, R.S. Infiltration and deep flow over sloping surfaces: Comparison of numerical and experimental results. *J. Hydrol.* **2009**, *374*, 30–42. [CrossRef]
27. Patin, J.; Mouche, E.; Ribolzi, O.; Chaplot, V.; Sengtaheuanghoung, O.; Latsachak, K.O.; Soulileuth, B.; Valentin, C. Analysis of runoff production at the plot scale during a long-term survey of a small agricultural catchment in Lao PDR. *J. Hydrol.* **2012**, *426–427*, 79–92. [CrossRef]
28. Mu, W.; Yu, F.; Li, C.; Xie, Y.; Tian, J.; Liu, J.; Zhao, N. Effects of rainfall intensity and slope gradient on runoff and soil moisture content on different growing stages of spring maize. *Water* **2015**, *7*, 2990–3008. [CrossRef]
29. Khan, M.N.; Gong, Y.; Hu, T.; Lal, R.; Zheng, J.; Justine, M.F.; Azhar, M.; Che, M.; Zhang, H. Effect of slope, rainfall intensity and mulch on erosion and infiltration under simulated rain on purple soil of south-western Sichuan province, China. *Water* **2016**, *8*, 528. [CrossRef]
30. Morbidelli, R.; Saltalippi, C.; Flammini, A.; Cifrodelli, M.; Picciafuoco, T.; Corradini, C.; Govindaraju, R.S. Laboratory investigation on the role of slope on infiltration over grassy soils. *J. Hydrol.* **2016**, *543*, 542–547. [CrossRef]
31. Wang, J.; Chen, L.; Yu, Z. Modeling rainfall infiltration on hillslopes using flux-concentration relation and time compression approximation. *J. Hydrol.* **2018**, *557*, 243–253. [CrossRef]
32. Melone, F.; Corradini, C.; Morbidelli, R.; Saltalippi, C. Laboratory experimental check of a conceptual model for infiltration under complex rainfall patterns. *Hydrol. Proc.* **2006**, *20*, 439–452. [CrossRef]
33. Melone, F.; Corradini, C.; Morbidelli, R.; Saltalippi, C.; Flammini, A. Comparison of theoretical and experimental soil moisture profiles under complex rainfall patterns. *J. Hydrol. Eng.* **2008**, *13*, 1170–1176. [CrossRef]
34. Morbidelli, R.; Corradini, C.; Saltalippi, C.; Govindaraju, R.S. Laboratory experimental investigation of infiltration by the run-on process. *J. Hydrol. Eng.* **2008**, *13*, 1187–1192. [CrossRef]
35. Morbidelli, R.; Saltalippi, C.; Flammini, A.; Cifrodelli, M.; Corradini, C. A laboratory experimental system for infiltration studies. *Hydrol. Res.* **2017**, *48*, 741–748. [CrossRef]
36. Morbidelli, R.; Saltalippi, C.; Flammini, A.; Rossi, E.; Corradini, C. Soil water content vertical profiles under natural conditions: Matching of experiments and simulations by a conceptual model. *Hydrol. Proc.* **2014**, *28*, 4732–4742. [CrossRef]

Article

Modelling Recharge from Irrigation Developments with a Perched Water Table and Deep Unsaturated Zone

Glen R. Walker [1],*, Dougal Currie [2] and Tony Smith [2]

[1] Grounded in Water, 2/490 Portrush Rd, St Georges, Adelaide, SA 5064, Australia

[2] CDM Smith, Level 2, 238 Angas Street, Adelaide, SA 5000, Australia; curriedr@cdmsmith.com (D.C.); tonsmi@iinet.net.au (T.S.)

* Correspondence: glen.walker@internode.on.net; Tel.: +61-0417-668824

Received: 2 March 2020; Accepted: 24 March 2020; Published: 26 March 2020

Abstract: Modelling of recharge under irrigation zones for input to groundwater modelling is important for assessment and management of environmental risks. Deep vadose zones, when coupled with perched water tables, affect the timing and magnitude of recharge. Despite the temporal and spatial complexities of irrigation areas; recharge in response to new developments can be modelled semi-analytically, with most outputs comparing well with numerical models. For parameter ranges relevant to the western Murray Basin in southern Australia, perching can reduce the magnitude of recharge relative to irrigation accessions and will cause significant time lags for changes to move through vadose zone. Recharge in the vicinity of existing developments was found to be similar to that far from existing developments. This allows superposition to be implemented spatially for new developments, thus simplifying estimation of recharge. Simplification is further aided by the use of exponential approximants for recharge responses from individual developments.

Keywords: hydrology; irrigation recharge; perched water tables; groundwater modelling; vadose zone

---

1. Introduction

Irrigated agriculture leads to greater infiltration of water than non-irrigated agriculture, especially in semi-arid and arid regions [1]. This, in turn, leads to greater recharge to underlying groundwater systems, where it can cause salinization and waterlogging of agricultural land and salinization of water resources [2] and associated riparian zones. Where groundwater is fresh, irrigation recharge may have environmental benefits in returning fresh groundwater to streams and groundwater-dependent ecosystems and providing recharge for groundwater users [3].

The assessment and management of irrigation environmental risks requires an understanding of the processes that link actions, such as irrigation development and subsequent water use measures, and the environmental impacts. This linkage is not just characterized by changes in water fluxes, but in time delays for pressure changes to move from the site of the action (irrigation fields) to the site of the impact (streams, affected land, groundwater-dependent ecosystems). Where water tables are deep, the unsaturated zone under irrigation areas is an important pathway between actions and groundwater systems, that link to impacts [4]. Yet, this zone is often poorly understood, falling between the disciplines of the agronomic engineering and hydrogeology, meaning that links between actions and impacts may be poorly understood.

Previous models of the unsaturated zone have mostly assumed that water moves under gravity towards the water tables [5,6]. This means that below the agricultural zone, fluxes do not change in magnitude from the agricultural soil zone to the groundwater. However, the larger soil water fluxes under irrigation may not be able to be transmitted by gravity [7], where soil vertical hydraulic conductivity is low. This leads to conditions of perched water tables, with increased hydraulic gradients, saturated conditions and lateral movement of water above these low-conductivity zones [8–10]. Sub-surface drainage may be required to avoid waterlogging and land salinity and water may be lost to the land surface from the perched layer by drainage and evapotranspiration. Perched water tables may, therefore, both change the magnitude of vertical fluxes and 'smear' the movement of pressure changes over time. Modelling perched water tables is important for recharge in karstic geology [11], managed aquifer recharge [12], contaminant sites [8–10] and ecology of ephemeral streams [13,14].

Irrigation districts can be complex, with different zones being developed over time over a range of soils and water table depths. Any unsaturated zone model needs to be used in conjunction with groundwater models, which are often implemented to assess irrigation risks and support the design of mitigation measures. Unsaturated zone models, therefore, need to represent the main processes in these complex irrigation districts that are important for the assessment and management of risks; yet be simple enough to be practically implemented. The unsaturated zone models are required to link management actions, such as water use measures to recharge across the irrigation district as a function of time, and space.

While numerical hydrological models are able to model these processes, the implementation of such models under complex irrigation districts with perched water tables is often impractical. Simplicity of process representation may be addressed by keeping the number of parameters small, seeking data sources that may be used for the parameterization and calibration of parameters, and using algorithms that can be run relatively quickly in conjunction with the groundwater model. The models and algorithms should be capable of representing the lifetime of irrigation districts, including new developments, water-use efficiency measures and decommissioning.

This paper describes the semi-analytical PerTy3 model, which has been developed to address issues in the western Murray Basin in southern Australia. Basin-wide strategies have been responsible for reducing river salinity in the lower reaches of the River Murray [15]. This has been possible through the combination of groundwater pumping and incentives for improving water use efficiency of irrigation districts. Groundwater models have been implemented across the western Murray Basin to assess salt load to the river. The highest-risk irrigation areas in the western Murray Basin overlie a saline regional groundwater system, which discharges into the River Murray. The paper tests and documents the PerTy3 model, as it relates to new developments.

However, even the application of semi-analytical models can be complex and resource-intensive, so this paper also explores the application of conceptual transfer functions [16–18] for individual actions, that can be superimposed. If such functions are shown to work, it would allow a transfer function model to be used for each development and action and then added, thus simplifying the estimation of recharge. Such models have previously been used for regions in which there are frequent fluctuations of the water table in response to recharge events.

Finally, since perched water tables lead to lateral movement of water, there is a need to consider both 'greenfield' and 'brownfield' developments. A greenfield development is one for which there is no hydraulic interference from nearby irrigation areas, while the opposite is true for brownfield developments. The antecedent moisture caused by prior nearby developments is thought to reduce time lags for wetting fronts to reach the water table. If so, there would be a need to consider the configuration of irrigation developments, which would add considerable complexity to the estimation of recharge under irrigation areas. This paper will consider the effect of interference between irrigation fields.

The aims of this study are to:

1. develop and test a semi-analytical unsaturated zone model to predict recharge under new irrigation developments with perched water tables. The main attributes being sought are:

    (a) a conceptual model representing the main physical processes at the scale of the irrigation district and periods of seasons and years;

    (b) continuity of modelling between conditions of perching and non-perching;

(c)   a limited number of additional parameters;

(d)   availability of field data that could be used to calibrate parameters;

(e)   benchmarking against appropriate numerical models;

(f)   use of agronomic modelling outputs as input and generates recharge as output to be used in groundwater models; and

(g)   a process for estimating recharge under brownfield developments.

2.   explore the use of even simpler modelling approaches based upon:

(a)   superposition of actions; and

(b)   conceptual approximants for transfer functions.

A further paper [19] describes the development of models for change in recharge due to water-use efficiency measures and for the whole-of-lifetime irrigation, including new development, water-use measures and decommissioning. A pilot study for the Loxton-Bookpurnong district in the western Murray Basin is described in [20]. The soil properties and other characteristics of that area will be used within this paper.

2. Theory

This section describes the processes and theory that underlies the semi-analytical model, PerTy3 and perched water tables, more generally. More specifically, this section supports objectives 1(a–c); namely, the description of a conceptual model representing the main physical processes at the scale of the irrigation district and periods of seasons and years; provides continuity of modelling between conditions of perching and non-perching; and includes a limited number of additional parameters. Non-dimensionalisation is used to group related parameters and, therefore, simplify the model.

The motivation for this study is illustrated in Figure 1, which shows the hydrogeological model of the Loxton-Bookpurnong irrigation district in southern Australia. Irrigation development has led to a perched water table above a low-conductivity clay layer and a groundwater mound in the underlying regional groundwater system. The increased groundwater gradients lead to greater volumes of saline groundwater entering the River Murray.

Figure 2 shows the conceptual model for the fluxes in the vadose zone. There are three semi-infinite layers of homogeneous soils, in which the second layer is of lower permeability. The left boundary condition is a no flow boundary, which means this is a line of symmetry. The upper boundary is the base of the agronomic zone, the left side of which underlies irrigated agriculture, and to the right side underlies non-irrigated agriculture. The upper boundary condition is a downward water flux, irrigation accession, as determined by agricultural practices, including channel leakage and spillage. For the one-dimensional systems discussed below, the whole upper surface is irrigated. The lower boundary condition is the water table, which is assumed to be constant. The profile is assumed to be initially at steady-state, with the root zone drainage at the boundary condition being relevant to either native vegetation or non-irrigated agriculture. At time zero, irrigation is implemented leading to an increase in root zone drainage. As the vertical flux may also consist of leakage or spillage from channels, it will be referred to as irrigation accession. In the western Murray Basin context, the new irrigation accession rate is ~100–400 mm/year), while the pre-irrigation flux (~0.1 mm/year for native vegetation or 2–30 mm/year for dryland agriculture).

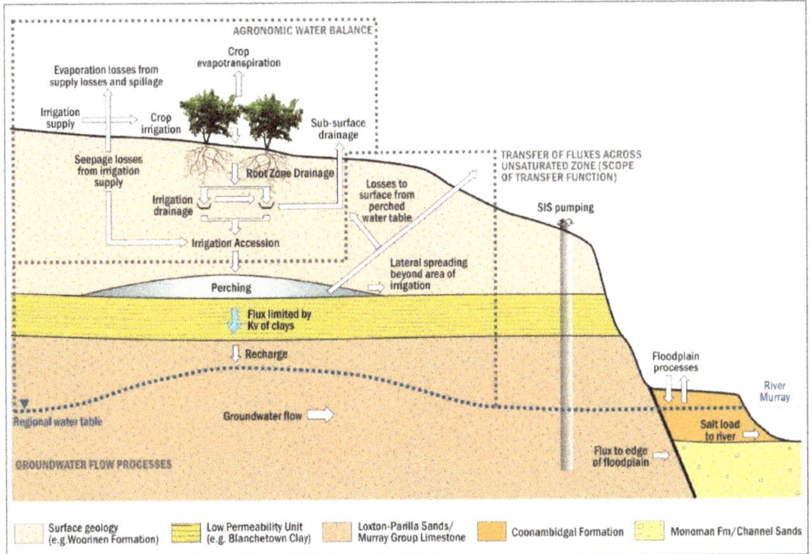

Figure 1. Conceptualizations of the Loxton-Bookpurnong Irrigation District, showing perched water table under the irrigation district, and groundwater flow to the River Murray. model used to simulate recharge under perched water tables.

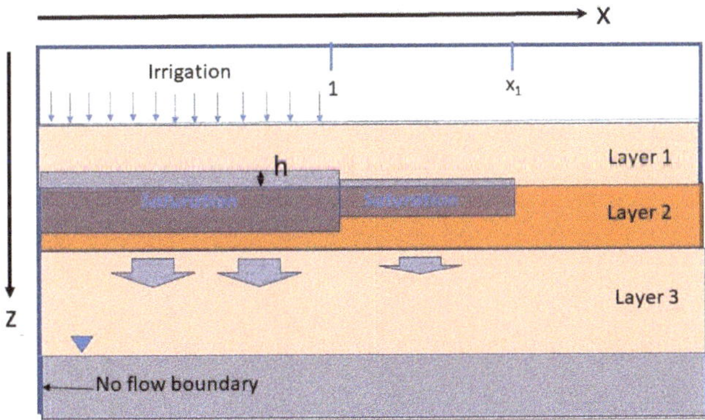

Figure 2. Used to simulate recharge under perched water tables. The left-hand boundary is a no flow boundary, representing a line of symmetry. The variables are non-dimensionalised, with $x = 1$ being the outer limit of irrigation and $x = x_1$ being the outer limit of perched water. Below layer 3 is the saturated zone of the aquifer.

This paper considers situations, where there is a reasonable probability of perched water tables under the irrigation area. This means that the irrigation accession should be sufficiently high or the saturated vertical conductivity sufficiently low for perched water tables to occur. Where it does occur, the ponded head builds up on the impeding layer and water moves laterally over the impeding layer, where it infiltrates into the impeding layer.

The layers of the unsaturated zone have been parameterised using the modified Mualem–Brooks–Corey model [21,22] for each layer. The water retention curve is given by:

$$\theta = (\psi/h_b)^{-\lambda}, \psi > h_b \tag{1}$$

$$\Theta = 1, \psi \leq h_b \tag{2}$$

where ψ is the soil suction, h_b is the air-entry point, λ is a fitting parameter, and the relative saturation, Θ, is given by

$$\Theta = \frac{(\theta - \theta_r)}{(\theta_s - \theta_r)}, \tag{3}$$

where θ is the volumetric water content, θ_r is the residual volumetric water content and θ_s is the saturation volumetric water content. The relative permeability, K_r, is given by:

$$K_r = \Theta^m, \tag{4}$$

where m is a fitting parameter related to the connectivity of soil pores.

The hydraulic conductivity, K, is obtained by multiplying K_r by the saturated hydraulic conductivity. These parameters will be different for the different layers and a subscript i will be subsequently used to distinguish layer i. The parameters are assumed to be the same for both vertical and horizontal properties, except for the saturated conductivity. A superscript h and v will be used with the saturated hydraulic conductivity to characterise anisotropy. The values given to the parameters is given in the Methods section.

The underlying equations are non-dimensionalised using the vertical length scale, l_2, timescale $S_2 l_2 / K_{s2}{}^v$, horizontal length scale x_0, and vertical flux $K_{s2}{}^v$; where l_i is the thickness of the ith layer; x_0 is the half-width of the irrigated area; $K_{si}{}^v$ is the saturated vertical conductivity of the ith layer; and S_2 is the specific yield for the 2nd layer for the initial dry conditions. The purpose of the non-dimensionalisation is to simplify the range of situations as much as possible using scaling and non-dimensional variables.

2.1. Unsaturated Zone Conditions

The PerTy3 model, as it relates to new developments, is adapted from the wetting front model [5]. In that model, the movement of water through the vadose zone occurs via gravity, causing a pressure (or wetting) front. Behind the wetting front, the flux of water is equal to the new flux, while below it, the flux equals the old flux. The wetting front moves with the speed (non-dimensional):

$$dz_{wf}/dt = (A_n - A_o)/(\theta_n - \theta_o) \tag{5}$$

where z_{wf} is the depth of the wetting front below the land surface; A_n and A_o are the non-dimensional irrigation accessions for the irrigated and pre-existing agriculture respectively; and θ_n and θ_o are the volumetric water contents above and below the wetting front. Their values are such that the relative vertical hydraulic conductivity equals A_n and A_o, respectively. When the pressure front reaches the water table, the recharge (dimensioned) increases from IA_o to IA_n. Equation (5) can be used to estimate the time delay between the change in land use and the change in groundwater recharge. The above theory, or variants of it, has been used to estimate time delays for changes in non-irrigated agriculture to affect the underlying groundwater. In this paper, we look to adapt this model to the situation, where the soil conductivity of parts of the vadose zone is sufficiently low to not allow the new water flux to move vertically by gravity alone. The simplicity of the model is appropriate for our knowledge of soil properties and input fluxes over representative scales. The parameters in Equations (1)–(4) will change for each layer. A subscript 'i' will be used to denote these parameters for layer i.

2.2. Situations Where Perched Water Tables Occur

The conceptual model in PerTy3 considers five stages for the pressure front to move through to the water table and for the new recharge rate to be attained. These are:

Stage 1: the pressure front moves through first layer according to Equation (5).

Stage 2: the interface between layers 1 and 2 needs to become saturated for the perched situation to occur. A wetting front continues to move through layer 2 while the moisture content about the interface is increasing. However, the dimensionless vertical flux across a broad zone from above the interface to below the pressure front is reducing from A_n to A_0.

Stage 3: saturated conditions develop at the interfaces of the first and second layers. This causes a saturation front to move downward behind the wetting front into layer 2, while the perched layer begins to build up in layer 1. The zone between wetting and saturation fronts is near-saturated and can be broad. The perched layer causes the hydraulic gradient in the saturated zone to be greater than that of gravity alone. As the ponded head rises, the hydraulic gradient continues to increase and the flux behind the wetting front increases. Water begins to move laterally above the interface between layers one and two and begins to infiltrate into the impeding layer.

Stage 4: the wetting front has reached the base of layer 2 and begins to move through layer three. The surface of the perched water table continues to rise towards an equilibrium, as does the flux behind the wetting front. As the flux increases, the saturation zone moves slowly towards the base of layer 2.

Stage 5: the wetting front has reached the water table. The recharge rises from the old irrigation accession rate. The recharge continues to rise until the perched water table reaches a new equilibrium. This occurs when the increased gradient through the second layer and the increased area of infiltration means that the recharge is equal to the irrigation accession. Where layer one is sufficiently thin, water from the perched water table is intercepted either by evapotranspiration or sub-surface drainage. This prevents the recharge from reaching the irrigation accession flux.

In adapting the wetting front model to perched situations, the following changes are incurred:

- Darcy's Law is applied to the saturated zone in the upper clay layer. We assume that all of the hydraulic resistance is due to the clay and hence proportional to the thickness of the saturated layer, while the ponded head means a hydraulic gradient greater than one, purely due to gravity.
- The ponded head increases to the stage where the irrigation accession flux can pass through layer 2, or it fills all of layer 1.
- The perching results in a distribution of transit times for a change in irrigation accession rate to reach the water table.
- There may be a difference in magnitude between the irrigation accession and recharge, but only where some of the irrigation accession is returned to the land surface. If not, at equilibrium the recharge rate equals the irrigation accession rate.
- We have found it necessary to no longer consider the wetting front as a sharp transition from pre-development conditions to saturation. The existence of air-entry suction, below which hydraulic conditions the same as saturation occur and the near saturated zone means that the transition can be significant.
- As the wetting front moves through layer 3, the increasing vertical water flux at the base of layer 2 can lead to the wetting front moving more quickly as it moves to the water table.

The sections below describe these stages in more mathematical detail. Table 1 lists all the parameters used, their symbols and units and the equation number, where first used.

Table 1. Glossary of parameters, symbols and units used in equations and the equation number, where first used.

Parameter	Unit	Symbol	Equation First Used
Dimensioned parameters			
Thickness of layer i	cm	l_i	(6), (10)
Time scale associated with equilibration process of perched water table	year	t_s	(34)
Soil conductivity for layer i—dimensioned (current, saturated horizontal, saturated vertical)	cm/day	$K, K_{si}{}^h, K_{si}{}^v$	(10)
Irrigation accession flux (current, new, old, for step j)	mm/year	IA, IA_n, IA_o, IA_j	(11)
Half-width of the irrigated area	m	x_0	(10)
Air-entry potential	cm	h_b	(1)
Soil water suction (negative potential)	cm	ψ	(1)
Recharge to the water table (current, change)	mm/year	$R, \Delta R$	(41)
Sub-surface drainage	mm/year	D	(40)
Dimensionless parameters			
Depth of the wetting front (wf) or saturation front (sat) below the base of layer 1		z_{wf}, z_{sat}	(5)
Lateral distance from centre of irrigation field		x	(8)
Time (current, initial, for wetting front to reach base of layer 1, time for saturation to occur, time for wetting front to reach base of layer 2, time for which flux changes at base of layer 2)		$t, t_0, t_1, t_2, t_3, t_4$	(5), (6), (7), (45)
Soil volumetric water content for layer i (current, residual, saturated, new, old)		$\theta_i, \theta_{ri}, \theta_{si}, \theta_n, \theta_o$	(3), (5)
Relative permeability for layer i		K_{ri}	(3)
Ratio of specific yields for layers 1 and 2		β	(8)
Relative saturation		Θ	(1)
Specific yield for the ith layer for the initial dry conditions		Si	(6)
Mualem exponent		m_i	(4)
Coefficient for soil water retention curve		λ	(1)
Head of perched water table (current, initial, equilibrium, at edge of irrigation field)		h, h_0, h_{eq}, h_1	(13), (7), (32)
Dimensionless irrigation accession (current, new old)		A, A_n, A_o	(5)
Dimensionless parameters related to thickness of near-saturated zone		φ	(26)
Thickness of transitional zone between saturation front and wetting front		Φ	(26)
Dimensionless parameter related to the significance of lateral movement		B	(8)
Vertical water flux through impeding layer (current, equilibrium, old)		q, q_{eq}, q_0	(8)
Specific yield of layer 1, relative to soil following passage of wetting front		s_1	(35)
Width of wetted zone outside irrigation field (current, equilibrium, old)		x_1, x_{1eq}, x_{10}	(8), (18)
Transfer function for the recharge		TF	(41)
Proportionality constant between perched head and depth of saturation front during stage 3		α	(21)
Velocity through layer 3 (group velocity for pressure changes, wetting front velocity)		v_g, v_{wf}	(36), (37)
Dimensionless parameter depicting lateral movement		B	(8)
Fitted parameter for approximants		a	(45)

2.3. Stage 1

During stage 1, the rate of movement of the wetting front is given by Equation (5). The dimensionless time for the wetting front to reach the top of the clay layer is given by:

$$t_1 = l_1(\theta_{n1} - \theta_{o1})/((A_n - A_o)l_2) \tag{6}$$

The variable t_1 is sensitive to θ_{n1}, this needs to be calculated for the new flux.

2.4. Stage 2

Once the wetting front reaches the interface between layers 1 and 2, the low permeability of layer 2 means that moisture begins to accumulate about the interface. The additional moisture creates a moisture gradient both below and above the interface in different directions. These gradients allow more or less moisture than would occur by gravity alone at that moisture content and allows the flux and soil suction to be continuous at the interface. A mass balance argument means that the difference in flux at the upper and lower ends of the segment is equal to the rate of accumulation in that segment. The time for saturation to occur is given by:

$$t_2 = \int (\theta - \theta_b)dz/(A_n - A_o),\tag{7}$$

where θ_b is the background water content. For layer 1, this background value is the new moisture content and for layer 2, this is the old moisture content. We will return to the calculation of integral (7) later.

2.5. Stage 3

The dimensionless equations describing the mass balance of the perched layer are:

$$\beta\frac{\partial h}{\partial t} = B\left(\frac{\partial}{\partial x}\left(h\frac{\partial h}{\partial x}\right)\right) + A - q \; 0 < x < 1 \tag{8}$$

$$\beta\frac{\partial h}{\partial t} = B\left(\frac{\partial}{\partial x}\left(h\frac{\partial h}{\partial x}\right)\right) - q \; 1 < x < x_1 \tag{9}$$

where

$$B = K_{s1}{}^h l_2{}^2/(K_{s2}{}^v x_0{}^2) \tag{10}$$

$$A = IA_n/K_{s2}{}^v \tag{11}$$

$$\beta = (\theta_s{}^1 - \theta_n{}^1)/(\theta_s{}^2 - \theta_o{}^2) \tag{12}$$

h is the head of the perched water table, $x = 1$ represents the edge of the irrigation field; $x = x_1$ is the edge of the wetted zone outside of the irrigation field; x_0 is the half-width of the irrigated field; and q is the vertical flux into layer 2. Continuity in h and the flux of water (and therefore gradient in h) is assumed to occur at $x = 1$. At $x = x_1$, h is zero.

For the above equations, A is a dimensionless parameter that reflects the degree of perching, with perching not occurring for smaller A and interception of the perched head with the upper boundary condition for larger A. The dimensionless parameter, B, reflects the degree to which lateral movement occurs. As B approaches zero, there is no lateral movement and the system behaves as a 1D system. For very large B, the perched layer spreads thinly across the impeding layer. We also shall assume for this stage, that the head of the perched layer is lower than the upper boundary, i.e.,

$$h \leq l_1/l_2 \tag{13}$$

Darcy's Law across the saturated zone implies that:

$$q = 1 + h/z_{sat} \tag{14}$$

where z_{sat} is the depth of the saturation front. This assumes that the main hydraulic impedance is in the second layer. Under the wetting front model,

$$dz_{wf}/dt = q \tag{15}$$

This assumes that the new flux is much greater than the old flux. During Stage 3, the wetting front has not reached the base of layer 2.

In addition to the equations above, some further assumptions are added in order to estimate recharge:

1. We shall ignore the effect of the ponded head outside the irrigated field on the infiltration into the impeding layer, i.e.,

$$q = dz_{wf}/dt = 1, 1 < x < x_1 \tag{16}$$

2. We shall assume quasi-steady-state Depuit–Forchheimer equations for this area, which leads to the following equations:

$$h = (1 + sqrt(B)h_1 - x)/sqrt(B), \quad 1 < x < x_1 \tag{17}$$

$$x_1 = h_1 sqrt(B) + 1, \text{ and} \tag{18}$$

$$Q = -Bh\frac{\partial h}{\partial x} = h_1 sqrt(B) \tag{19}$$

where Q is the non-dimensional lateral flux at $x = 1$ and h_1 is the head of the perched layer at $x = 1$.

3. We shall assume that the head is constant across the irrigated field. Combining Equations (8), (9), (18) and (19) gives:

$$\beta dh_1/dt = A - q - h_1 sqrt(B). \, 0 < x < x_1 \tag{20}$$

4. We shall assume in early times of ponding that the lateral movement is small, and processes are vertical. We shall also assume that the separation between saturation fronts and wetting fronts is constant. By defining the dimensionless parameter:

$$\alpha = h/z_{sat,} \tag{21}$$

we find that z_{sat} and h increase linearly:

$$z_{sat} = (1 + \alpha)t \tag{22}$$

$$h = \alpha(1 + \alpha)t \tag{23}$$

$$\alpha = (-(1 + \beta) + sqrt((1 + \beta)^2 + 4(A - 1)\beta)/(2\beta), \text{ and} \tag{24}$$

$$q = 1 + \alpha \tag{25}$$

Equation (25) indicates a flux greater than the free drainage flux through a saturated clay layer ($q = 1$).

Stage 3 finishes when the wetting front reaches the bottom of layer 2. To estimate when this occurs, it is necessary to estimate the thickness of the zone, Φ, between the wetting front and the saturated zone (Figure 3a). If we assume that this zone is in a quasi-steady state, this can be estimated from Darcy equation to give:

$$\Phi = \varphi/(q - 1) = \int d\psi/((q/K_r(\psi) - 1)l_2), \tag{26}$$

$$\Phi = h_{b2}/((q - 1)l_2) + \int d\psi/((q/(K_r(\psi) - 1)l_2) \tag{27}$$

where φ is a soil hydraulic property and the integral in Equation (26) goes from 0 (saturated) to ψ_3, and in Equation (27) from h_{b2} to ψ_3, where ψ_3 is either (a) the matric potential relating to the pre-development drainage, where the transition zone is entirely within the clay layer or (b) the soil matric suction at the interface with layer 3. We shall assume that φ is constant with respect to q.

Once the wetting front reaches the base of layer 2, the flux at top of layer 3 increases to $q = 1 + \alpha$ and the wetting front begins to move through this layer. This occurs at non-dimensional time after perching begins:

$$t_3 = (1 - \varphi/\alpha)/(1 + \alpha), \tag{28}$$

and the ponded head will be:

$$h_o = \alpha - \varphi \tag{29}$$

By knowing the flux during this third stage, it is possible to use the steady-state Darcy's Law to calculate t_2 in Equation (7).

2.6. Stage 4

After the wetting front reaches the base of layer 2, Equations (14) and (26) can be combined to estimate the vertical flux:

$$q = h + 1 + \varphi \tag{30}$$

Incorporating Equation (30) into (20), the mass balance for the perched layer becomes:

$$\beta\frac{\partial h}{\partial t} = A - 1 - \varphi - h \left(1 + sqrt(B)\right) \tag{31}$$

where φ is calculated using Equation (27). This has the solution:

$$h = h_{eq} + exp(-(t - t_0)(1 + sqrt(B))/\beta)(h_0 - h_{eq}) \tag{32}$$

where h_0 and t_0 are, respectively, the head and time at which the wetting front breaks through the clay layer Equation (29). The equilibrium head, h_{eq}, is given by:

$$h_{eq} = (A - 1 - \varphi)/(1 + sqrt(B)) \tag{33}$$

The effect of B is to not only reduce the steady-state ponded head, but it also quickens the rate at which it is attained. To understand this better, we consider the dimensioned time scale:

$$t_s = l_2 S_2 \beta/((1 + sqrt(B)) K_{s2}{}^v) \tag{34}$$

As B becomes very large, t_s becomes:

$$t_s \sim x_0 s_1/sqrt(K_{s1}{}^h\, K_{s2}{}^v) \tag{35}$$

Hence, the time scale involves a mixture of the horizontal conductivity of the sand layer and the vertical conductivity of the clay layer. This is not surprising given that both soil parameters influence both the magnitude of the ponded head and the ability of the water to move laterally.

Equation (30) implies that the vertical flux through the clay layer would also approach exponentially the equilibrium value q_{eq} from q_0. Equation (30) and Equation (32). The extent of the wetting outside the irrigation field also increases exponentially to the equilibrium value x_{1eq} from x_{10} in parallel with the ponded head (Equation (18) and Equation (32)). Equation (16) implies that the aggregated vertical flux through the clay external to the irrigation field is proportional to the extent of wetting.

As the flux at the base of layer 2 increases, the velocity of the wetting front can increase. The speed of the wetting front is given by Equation (5), but with A replaced by q. Unlike the situation where there is no perching, the water content at the wetting front, θ_{wf}, is likely to change as the flux at the bottom of layer 2 changes gradually. The change in flux (and associated water content) will move at a speed determined by the group velocity, $dK/d\theta$. This will continue until the change reaches either the wetting front or the capillary fringe (in the case, where the wetting front has already reached the capillary fringe). In the former situation, the time at which the change reaches the wetting front is given by:

$$t - t_4 = z_{wf}/(dK/d\theta(\theta(t_0))) = z_{wf}/v_g, \tag{36}$$

where t_4 is the time at which the change at the change occurs at the bottom of layer 2 and $dK/d\theta(\theta(t_4)) = v_g$ is the group velocity as determined there.

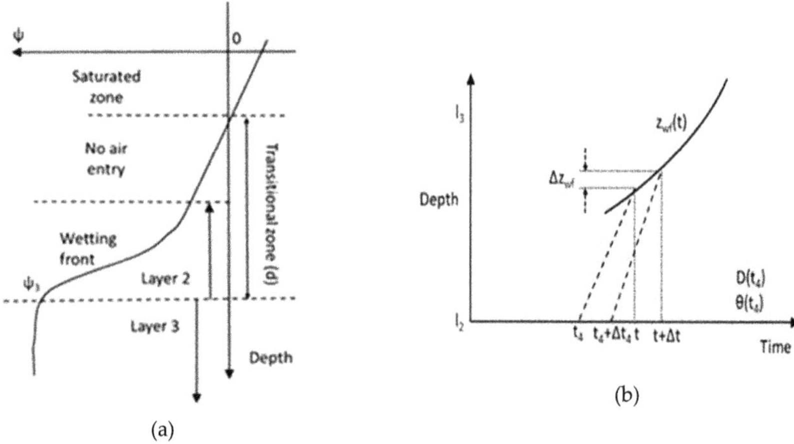

Figure 3. Figures showing two aspects of the modelling: (**a**) near-saturated zone between wetting front and saturation front, as described in Equation (26); and (**b**) changing speed of wetting front, as flux behind increases.

Figure 3b shows the process of calculating the rate of movement of the wetting front. Following the logic of that diagram allows z_{wf} to be calculated by integrating the following equation:

$$\Delta z_{wf} = v_{wf}\Delta t_4/(1 + v_{wf}/v_g) \tag{37}$$

where v_{wf} is the velocity of the wetting front and v_g is the group velocity. If $dK/d\theta \gg \Delta K/\Delta\theta$, as is the case for our parameterisation of layer 3, this simplifies to:

$$z_{wf} = \int v_{wf}dt_4 \tag{38}$$

Equation (38) implies that changes in flux and water content at the base of layer 2 are transmitted quickly to the wetting front. This causes the wetting front to move more quickly. Once the wetting front reaches the water table, the changes in flux should transmit quickly to the water table. Under the assumption that $dK/d\theta \gg \Delta K/\Delta\theta$, the gap in time between the flux at the base of layer 2 disappears. The effect of the higher fluxes at the base of layer 2 during the propagation of the wetting front through layer 3 is to quicken the pace of the wetting front but also to increase the flux at the time the wetting front reaches the water table. The speed at which the wetting front moves through layer 3 is less than that for the unperched situation as the flux at the base of layer 2 is less than for the unperched situation.

Computationally, the inclusion of the assumption simplifies and quickens the algorithms but not including the assumption is still computationally viable. The simpler assumption is made in PerTy3.

2.7. Stage 5

Once the wetting front reaches the water table, recharge rises from the pre-development flux to a new higher flux, less than the irrigation flux. The recharge rate continues to increase, exponentially asymptoting to the irrigation accession flux. The program ignores the effect of capillary fringe on timing. For example, for native vegetation or dryland agriculture the corresponding soil suction could be up to 10 m, while for irrigated agriculture, the corresponding suction is more likely 10's of cm. We shall assume that the bottom of layer 3 is the capillary fringe corresponding to irrigation.

If the perched head rises to the stage where it intercepts the upper boundary condition, there is no capacity for further infiltration to occur. We refer to this as recharge rejection. This occurs when:

$$h_{eq} = (A - 1 - \varphi)/(1 + sqrt(B)), h \geq l_1/l_2 \tag{39}$$

Any excess water is returned to the surface as evapotranspiration. This process can lead to waterlogging and salinity and generally sub-surface drainage is required. The volume of rejected recharge can be estimated using Equations (19) and (30):

$$D/IA = (A - 1 - \varphi - h(1 + sqrt(B)))/A \tag{40}$$

In portraying outputs, a normalised transfer function is often used:

$$TF(t) = (R(t) - Ro)/(IA_n - IA_o) \tag{41}$$

A transfer function is a mathematical model of a system that maps its input to its output (or response). An analogous transfer function can be defined for drainage (or rejected recharge). Where there is no rejected recharge, $TF(t) = 0$ for $t = 0$; and approaches 1 after long periods. Hence it represents a cumulative probability distribution for time delays for pressure to travel through the vadose zone. Where there is rejected recharge, $TF(t)$ is less than one after long periods but the sum of transfer functions for recharge and drainage approaches one. Transfer functions take on greater significance where they can be superimposed for a combination of actions. A modified transfer function will be defined below for application to superposition.

2.8. Theory: Summary

1. Irrigation development leads to the formation of a wetting front that moves through layer 1.
2. Once the wetting front reaches the interface between layers 1 and 2, the low permeability of layer 2 means that moisture begins to accumulate about the interface.
3. The ponded head is zero until the end of Stage 2, increases linearly until end of stage 3 and then exponentially asymptotes to an equilibrium head during stages 4 and 5.
4. The flux at the base of layer 2 is zero until the end of stage 3 and then increases exponentially to the irrigation accession rate.
5. The recharge rate is zero until the end of stage 4 and then increases exponentially to the irrigation accession flux.
6. While the ponded head increases to the stage where the irrigation accession flux passes through layer 2, the value of this may be so high that it intercepts the upper surface.

3. Methods

This section describes the modelling methodology used to meet objectives 1 and 2(a–b) of the study; namely to benchmark the PerTy3 model against an appropriate numerical model; develop a process for estimating recharge under brownfield developments; and explore the use of even simpler modelling approaches based upon superposition of actions, and conceptual approximants for transfer functions.

3.1. Model Implementation

To achieve these objectives, a series of implementations of the PerTy3 and the numerical model, FEFLOW (finite-element subsurface flow simulation system), are used. The default parameters used for both the one- and two-dimensional (2D) modelling are shown in Table 2. One-dimensional (1D) situations represents those, where lateral movement of water is minimal, and hence where B approximates zero. The default parameters have been derived using published estimates of soil hydraulic parameters for the Mallee region [23] and defining equivalent parameters for the Brooks–Corey–Mualem model. The irrigation flux pre-development, IA_o, has been assumed to be 10 mm/year for all experiments. For the two-dimensional experiments, the half-width of the irrigation area is assumed to be 500 m. Pertinent information on the irrigation water balance is from [24].

The numerical modelling is undertaken using FEFLOWTM [25]. FEFLOW solves the governing flow equations in porous media for variable saturation. Richards' equation is solved for a single dominant fluid phase (in this case water) with an assumed stagnant air phase that is at atmospheric pressure everywhere. FEFLOW implements a number of empirical and spline models for variable saturation. Fine mesh refinement is used to achieve stable numerical solutions for the adopted choices of the Mualem–Brooks–Corey model parameters. The vertical mesh size for both the 1D and 2D modelling is 10 cm (250 (1D) or 150 (2D) elements), while for the 2D modelling, the horizontal mesh size is 10 m (200 elements).

Table 2. Default soil parameters used in the modelling.

Parameter	Symbol	Layer 1	Layer 2	Layer 3
Texture		Sandy Loam	Clay	Sand
Saturated volumetric water content (cm^3/cm^3)	θ_{si}	0.35	0.4	0.38
Residual water content (cm^3/cm^3)	θ_{ri}	0.03	0.1	0.04
Air-entry potential (cm)	h_{bi}	12.0	40.0	8.0
Mualem exponent	m_i	8.24	7	6.94
Vertical saturated hydraulic conductivity (cm/day)	K_{si}^{v}	300	various	500
Anisotropy for saturated conductivity (Horizontal/Vertical)		~0 (1D) various (2D)	~0 (1D) 1(2D)	~0(1D) 1 (2D)
Thickness (cm)	l_i	500	500	1500 (1D) 500 (2D)

Table 3 details the various modelling experiments and the associated parameters. The modelling has been designed to achieve the various objectives:

1. 1–6 (1D), 1–4 (2D) are a series of simulations for new developments that cover a range of non-perched and perched situations and illustrate varying degrees of lateral movement. These simulations demonstrate the main processes and the outputs allow benchmarking of the models. For each experiment, the PerTy3 and FEFLOW models are used.

2. The experiments 5–10 (2D) are designed, in conjunction with Experiments 1–4 (2D), to explore the effect of a brownfield developments in the vicinity of a development, already at equilibrium. More specifically, the recharge under a greenfield development (1–4(2D)) is compared to brownfield developments either directly adjacent to or 250 m away from a development at equilibrium. The FEFLOW outputs will be compared to the superposition of the two developments.

3. Experiments 10–13 (2D) are designed to compare recharge brownfield sites in the vicinity of greenfield sites, that are 5 years old. The FEFLOW outputs will be compared to the superposition of the two developments.

Table 3. Benchmarking experiments.

Model Expt Number	Vertical Saturated Hydraulic Conductivity (cm/day)	Anisotropy for Saturated Conductivity (Horizontal/Vertical)	New Irrigation Accession (mm/year)	Parameter	Parameter	Separation from Existing Development (m)
	K^v_{si}		IA_n	A	B	
1 (1D)	0.0913	~0	100	0.3	~0	
2 (1D)	0.0365	~0	100	0.75	~0	
3 (1D)	0.0183	~0	100	1.5	~0	
4 (1D)	0.00685	~0	100	4.0	~0	
5 (1D)	0.146	~0	400	0.75	~0	
6 (1D)	0.067	~0	400	1.5	~0	
1 (2D)	0.03	1000	200	1.83	0.1	
2 (2D)	0.03	10,000	200	1.83	1	
3 (2D)	0.03	100,000	400	3.65	10	
3a (2D)	0.03	100,000	200	1.83	10	
4 (2D)	0.03	10,000	100	0.91	1	
5 (2D) *	0.03	1000	200	1.83	0.1	0
6(2D) *	0.03	10,000	200	1.83	1	0
7(2D) *	0.03	100,000	200	1.83	10	0
8(2D) *	0.03	1000	200	1.83	0.1	250 m from equilibrium development
9(2D) *	0.03	10,000	200	1.83	1	"
10(2D) *	0.03	100,000	200	1.83	10	"
11(2D) *	0.03	1000	200	1.83	0.1	250 m from 5-year-old development
12(2D) *	0.03	10,000	200	1.83	1.0	"
13(2D) *	0.03	100,000	200	1.83	10.0	"

* experiments where no PerTy3 modelling was performed, but FEFLOW (finite-element subsurface flow simulation system) modelling and superposition experiments were performed.

3.2. Transfer Functions and Superposition

In this section, we describe the concept of a transfer function for application to superposition, to support objective 2(a). For this purpose, a modified transfer function for the recharge is defined:

$$TF'(t) = (R(t) - Ro)/(IA_n{}^* - IA_o{}^*) \tag{42}$$

where the superscript * indicates the minimum of IA and the maximum irrigation accession without rejected recharge. We can define a similar transfer function for drainage, TF'', where:

$$TF''(t) = (D(t) - D_0)/(IA_n{}^{**} - IA_o{}^{**}) \tag{43}$$

where D_0 is the original drainage rate, IA^{**} is the maximum of IA and the maximum irrigation accession that occurs without rejected recharge.

The general aim is to generate outputs, with appropriate accuracy, for a range of inputs. In line with information theory, we would look to see whether simplifications are possible, by using, for example, processes, such as superposition. As the system is not necessarily linear, there is no reason to believe that superposition would necessarily apply to a range of situations. Some situations are not expected to follow superposition including those where thresholds are involved, such as the initiation of perching or rejected recharge. The addition of two actions, that do not individually exceed the threshold but in aggregation do so, is not linear.

If superposition does apply, the aggregate transfer function for a sequence of actions that affect irrigation accession:

$$TF'(t) = (\sum_j (IA^*_{j+1} - IA^*_j)\, TF'_{j+1})/(IA^*_{p+1} - IA^*_0) \tag{44}$$

where IA_j is a sequence of modified irrigation accessions that occur from $j = 0$ to $j = p + 1$ and TF'_{j+1} is the modified transfer function that applies for a change of irrigation accession from IA^*_j to IA^*_{j+1}.

The pattern of the aggregate TF' with time would broadly follow that of IA, perhaps with time delays and some 'smearing' and capping at the maximum irrigation accession for which there is no rejected recharge. Where superposition applies, this could simplify the numerical process. The transfer function could be generated for individual actions alone and this allows the overall aggregated transfer function to be generated. While the theory above describes physical processes, the analytical model is simplified, with assumptions such as spatial homogeneity, flat surface elevations etc. It is possible to consider transfer functions as conceptual models, using enough complexity to broadly replicate the recharge response. Such models have previously been used for recharge [16,18].

For this paper, we will compare the recharge under a brownfield development and the original greenfield development with the superposition of the two individual recharge outputs. Brownfield sites were thought previously to have shorter time delays, as they have already been wetted. Should the superposition be a good approximation, this may simplify the estimation of recharge under a complex irrigation district by allowing each development to be considered individually and then aggregated.

3.3. Approximants

In this section, we explore the application of approximants for the transfer function. In general, an approximant is a function, series, or other expression which is an approximation to the solution of a problem. Here, we trial the application of a conceptual model, specifically a linear reservoir model with a time delay, to approximate transfer functions, namely:

$$TF(t) \sim 1 - exp(-a(\mathrm{t} - t_6)),\ t > t_5 \tag{45}$$

where a, t_5 and t_6 are fitted parameters. Such models have previously been used [18] for recharge through a deep vadose zone. Such conceptual models have been used widely in surface hydrology to calibrate surface flow models. The linear reservoir model forms a good approximation where the output (recharge) is a linear function of the storage (the mass of water in the unsaturated zone) [18].

4. Results

4.1. One-Dimensional Modelling

Figure 4a compares the outputs for FEFLOW and PerTy3 for transfer functions for modelling Experiments 1(1D)–6(1D). There was no perching for Experiments 1, 2 and 5. These resemble step-functions, with consistency of results between the two forms of modelling. The values of A for these experiments were 0.3, 0.75 and 0.75, respectively. The model outputs were again consistent, with Experiments 3 and 6 showing perching. The transfer function showed a step increase, followed by an apparent exponential approach to one. The model outputs for the ponded head for these experiments are also consistent (Figure 4b) and both show an exponential approach to equilibrium after a delay and an initial linear rise. The values of A for these experiments were both 1.5. The modelling outputs for Experiment 4 show the least consistency with the numerical rise in the transfer function being slower; and the ponded rise occurring slightly later but both intercepting the upper surface (500 cm). Drainage or rejected recharge occurs with the increase for the analytical function occurring nearly 5 years later than the numerical output for drainage. Overall, the modelling outputs show that the processes are correct and the outputs are adequate for use in groundwater modelling.

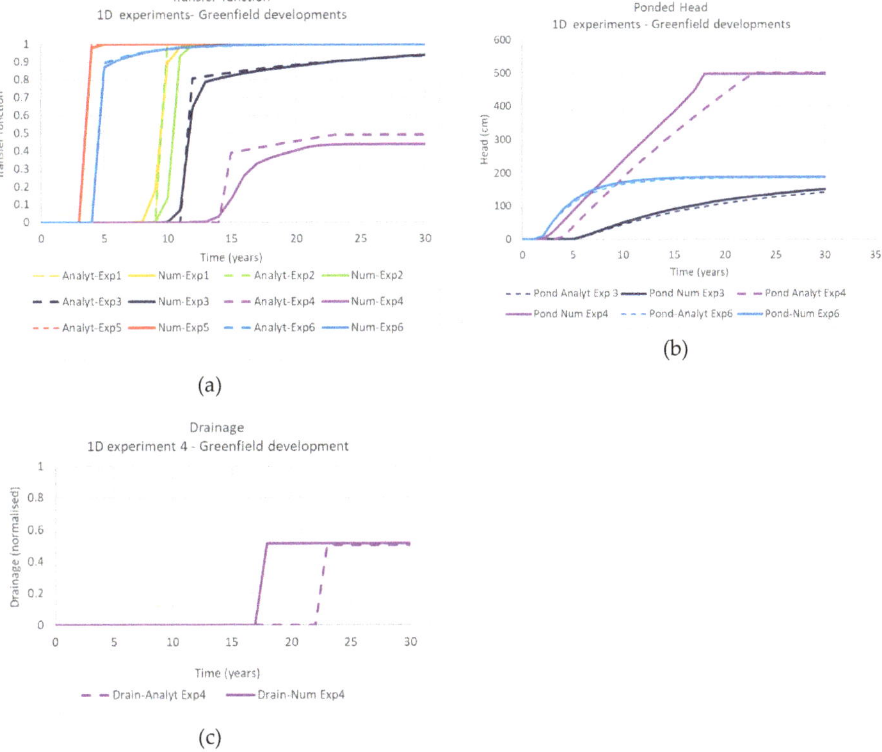

Figure 4. Of outputs from one-dimensional outputs from PerTy3 (semi-analyt-ical) (dashed) and FEFLOW (numerical) (solid) models (**a**) transfer functions; (**b**) ponded head and (**c**) normalized drainage volume.

4.2. Two-Dimensional Modelling

The outputs for both the FEFLOW and PerTy3 models are consistent across a range of values of B from 0.1 to 10 (Figure 5a). The PerTy3 outputs are flatter and lower than those from FEFLOW. This is due to the representation of recharge external to the irrigated field (Figure 5b). The partitioning between recharge occurring outside and inside the irrigated field is consistent across the models, but PerTy3 shows initial delays in recharge occurring outside the field. Another modelling experiment (not shown here) shows even greater inconsistency, suggesting that assumptions were not adequate. The most likely cause is the lack of consideration of ponding external to the irrigated field. This would have the effect of delaying recharge increases.

The FEFLOW output for the perched head (Figure 5c) if for $x = 0$, while that for PerTy3 is an average across the irrigation field. While the FEFLOW output is higher than the PerTy3, the temporal distributions are approximately parallel. There is a large change in head for $B = 0.1$ (340 cm) to $B = 10$ (120 cm). The wetted width PerTy3 outputs (Figure 5d) shows a large variation from 90 m ($B = 0.1$) to 250 m ($B = 10$).

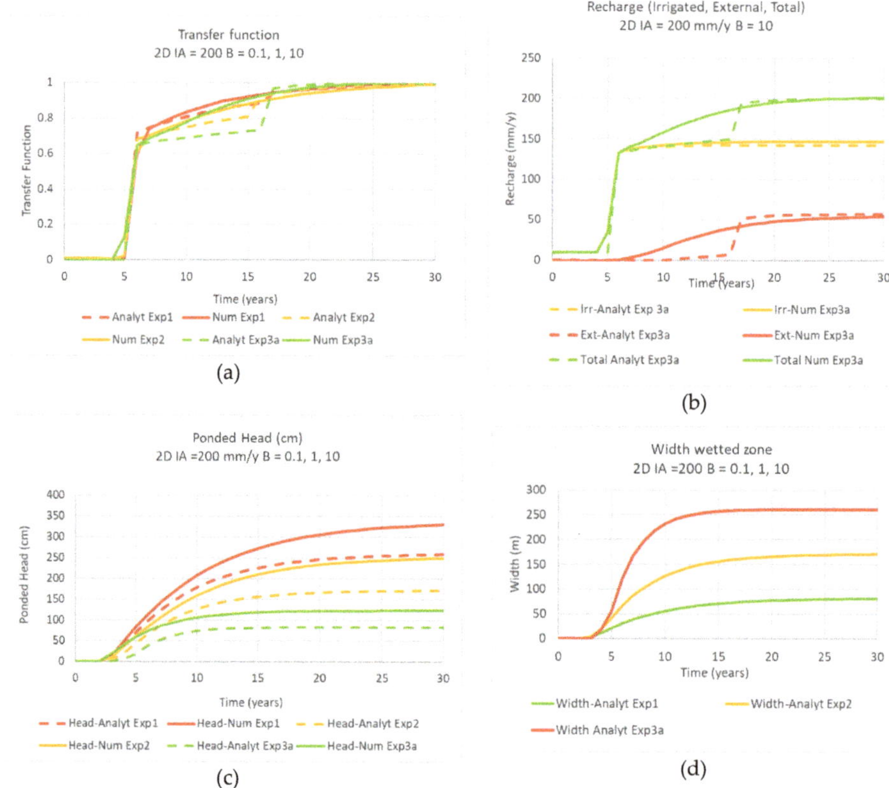

Figure 5. Comparison of outputs from two-dimensional semi-analytical and numerical modelling for new accession rate of 200 mm/year: (**a**) Transfer functions (Total) for $B = 0.1, 1, 10$; (**b**) Transfer functions (total, under irrigation field, external to irrigation field) for $B = 10$. (**c**) Perched heads for $B = 0.1, 1.0, 10$ for an increase in IA to 200 mm/year. The head for the FEFLOW model is for $x = 0$, while that for PerTy3 is an average across the irrigated field. (**d**) the width of the wetted zone outside the irrigation field for $B = 0.1, 1.0, 10$, and an increase in IA to 200 mm/year.

4.3. Modelling of Brownfield Developments

The effect of brownfield developments at a range of distances is shown in Figure 6. For low values of B (Figure 6a), there is almost no difference whether the new development is placed next to an existing field or at infinite distance (greenfield development). Even for B of 10, there is not much difference. For brownfield developments next to an existing development at equilibrium (Figure 6b), there is some earlier recharge and then later some delayed recharge. The earlier recharge is presumably due to pre-wetting by the existing development and the later delays due to the expansion on one side being constrained by the existing development.

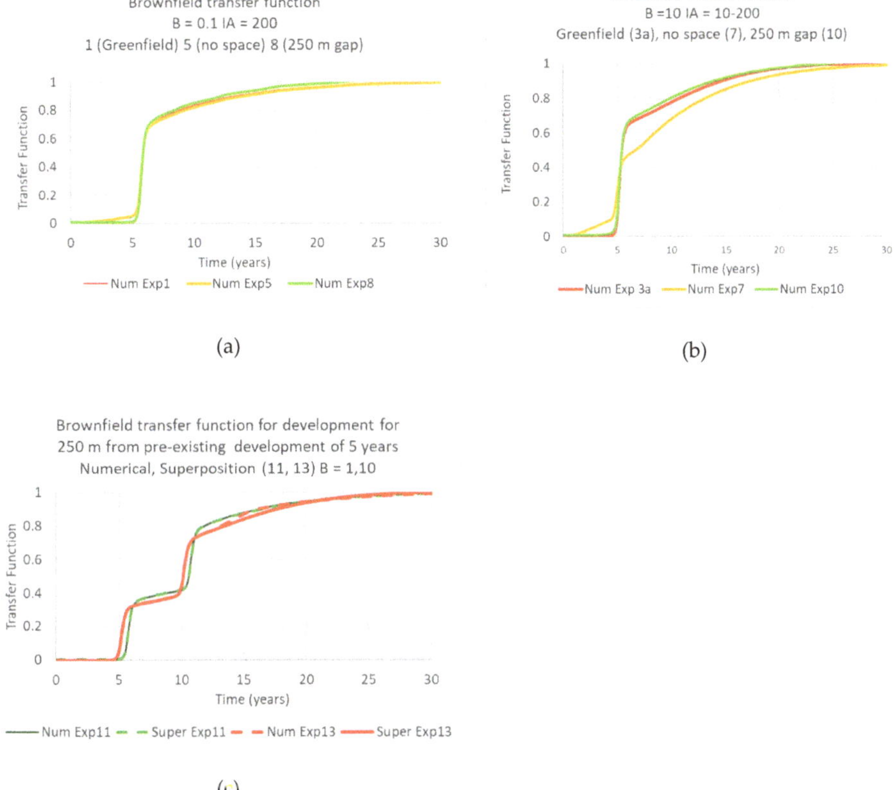

Figure 6. Functions for brownfield developments: (**a**) numerical outputs at varying distances from a pre-existing steady-state development and $B = 0.1$; (**b**) numerical outputs at varying distances from a pre-existing steady-state development and $B = 10$; (**c**) numerical outputs for the total development of a new development followed by another development, 250 m away for $B = 1, 10$ compared to superposition of numerical outputs for two independent developments, one 5 years after the other.

For brownfield developments occurring 5 years after a new development (Figure 6c) there is, minimal effect of separation.

4.4. Approximants

The trial of approximants was mostly successful with good matching with an exponential function (Figure 7). The worst fit was for Experiment 4 (1D) (Figure 7a). For this experiment, the ponded head reached the upper boundary condition by about year 17. This appears to change the temporal pattern of recharge, which is not captured by the approximant.

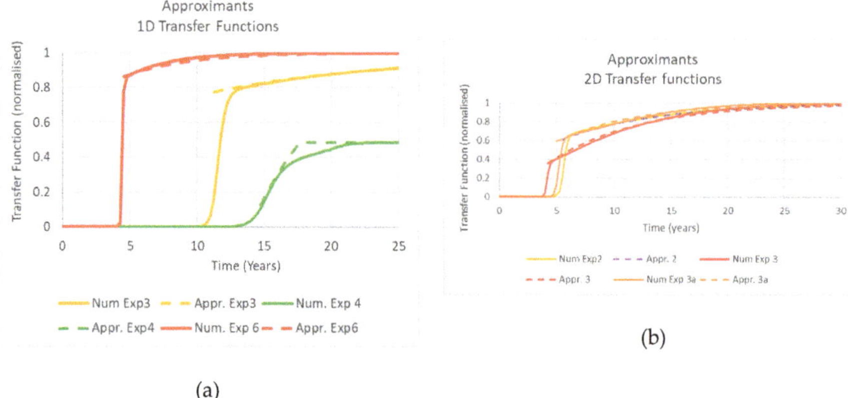

(a)

(b)

Figure 7. Approximants against numerical solutions for which perching occurs. Solid lines represent numerical solutions and dashed lines approximants. (**a**) 1D modelling experiments 3, 4 and 6. The approximants are respectively: $1 - \exp(-0.07 \times (t + 10))$, $t < 12$; 4: $\min(0.486, 1 - \exp(-0.18 \times (t - 14)))$, $t > 15$; and 6: $1 - \exp(-0.22 \times (t + 4.4))$, $t > 5$. (**b**) 2D Modelling Experiments 2, 3 and 3a. The approximants are respectively: 2: $1 - \exp(-0.12 \times (t + 2.5))$, $t > 6$; 3: $1 - \exp(-0.13 \times (t - 0.8))$, $t > 4$; and 3a: $1 - \exp(-0.15 \times (t + 1))$, $t > 5$.

The coefficient in the exponential is similar for all the 2D modelling experiments. The theory is predicting change in the exponential for experiment 3a. On the other hand, the 1D modelling is showing considerable variation, which was not expected. However, Figure 4a shows that the analytical expressions, which have similar coefficients for all the experiments, adequately fit the numerical simulation. The 1D modelling is trying to fit over small variations of the transfer function and then becomes sensitive to issues such as numerical dispersion.

Perhaps the largest difficulties are 1) the estimation of the time delay until recharge occurs and 2) the reference time in the exponential. However, the analytical model appears to be adequately estimating time delays and becomes an issue of finding the simplest form of estimation.

Overall, the collection of results shows promise for using approximants, in addition to using a PerTy3 or numerical models.

5. Discussion

5.1. Accuracy of Modelling

Modelling will necessarily involve simplifications of reality. For example, (1) all surfaces are assumed to be topographically flat; and (2) layers are considered to be physically homogeneous with known properties. The values of hydraulic properties need to be considered to be conceptual in nature, rather than being true values. The study here has tried to incorporate the key processes. The comparison between models tests mathematical errors.

The comparison of model outputs show that the outputs were consistent in most aspects for the 1D modelling. Two exceptions were (1) the transfer function for Experiment 4; the situation involving interception of the ponded head with the upper boundary condition; and (2) the drainage volumes for Experiment 4. These discrepancies were related in that the PerTy3 output showed that the perched head reached the upper boundary condition about five years later than the FEFLOW output. This has led to drainage volumes occurring five years later, although volumes were similar. It also meant that recharge has equilibrated at about the same time as this interception for PerTy3; while recharge has equilibrated some years after the perched head has intercepted the upper surface, causing a distortion of the transfer function. Despite these discrepancies, the consistency of the model shows that the representation of the processes are accurate.

The comparison of model outputs for 2D modelling does not show the same level of consistency. In particular, PerTy3, compared to FEFLOW, overpredicts (1) the delay in recharge occurring external to the irrigated field; and (2) the exponential decay in transfer function. Both may be due to ignoring the additional ponded head on recharge external to the irrigated field. The mound extends beyond the irrigated field [24] and this reduces delays for recharge to occur, causing recharge to increase continuously rather than be delayed. The mounding also changes the shape of the moisture storage in the unsaturated zone, which will affect the exponential decay [19].

The outputs for FEFLOW and PerTy3 2D models are consistent in the partitioning of recharge between under the irrigated field and external to the irrigated field; the reduction of the perched head with B, and the exponential decay of perched head and wetted width. Thus, while some 2-dimensional aspects are not adequately modelled, most other aspects appear to be adequate for the objectives of the transfer function.

The comparison of model outputs with field observations will be discussed in [20].

5.2. Brownfield Developments

There has been some discussion in the Australian context, that brownfield developments should respond more quickly due to antecedent moisture. The results here show that while this partially occurs for high values for B, there are no lateral impacts (i) for $B = 0.1$ or 1, or (ii) for separations of 250 m or more; and that (iii) for no separation, there are delays in recharge that outweigh the early breakthroughs. The lack of lateral impacts, especially for $B = 0.1, 1$, appears to be due to limited wetted extent outside irrigated fields. The later time delays for $B = 10$ are probably due to the inability of water to move laterally on one side. The effect would be similar to a larger irrigation field, with higher perched water table and relatively smaller wetted extent.

The small interaction between irrigated fields means that a complex irrigation area with new developments occurring at different times could be modelled as a superposition of impacts for individual fields. This would greatly simplify the calculation of recharge under the irrigation district. Should the circumstances be such that there are interactions (high B, small gaps between irrigation fields, the results here indicate that there may be simple approximations, but this would require more work.

5.3. Approximants

The study showed that the transfer functions for both 1D and 2D were able to be approximated by an exponential function with a reference time. The best-fit coefficient for the exponential was consistent across the 2D modelling, but variable for the 1D modelling. However, the analytical value was reasonable for five of the six 1D outputs, suggesting that one value may be a reasonable approximation for most situations. If so, this requires only for the reference time and the time delay (the time at which recharge, and the exponential approximation begins). This was not investigated in this study. However, PerTy3 appears to be able to estimate the time delay and the change in recharge at that time, so it may become a matter of finding how this could be simplified or how these parameters change in response to different model parameters. The exponential function can be derived from a linear reservoir model in which the total recharge is linear with respect to the mass of water in the perched water table.

The use of approximants and superposition may lead to simplification of the estimation of recharge. The approximants form a step towards considering transfer functions as conceptual models that are fitted by the groundwater response. This has been done previously [16,18] for situations where response times are quicker and individual events are independent. These are then used for calibration. For the situations considered here, the response from different actions gradually accumulate into one single evolving groundwater response. This can be used to calibrate the transfer function, where there is no rejected recharge and provided transfer functions do not vary greatly for different actions. Where there is rejected recharge, drainage can be used for calibration.

5.4. Data Requirements

Superficially, the model appears to require many parameters and hence should be considered as complex. For example, the model, in principle, requires several soil parameters for each layer. However, many of these parameters are based on texture and can be approximated based on lithology information and data [23]. The saturated vertical hydraulic conductivity for layer 2 and horizontal conductivity for layer 1 appear to be critical parameters and are reflected in the dimensionless variables A and B, which portray the range of behavior expected in the field [19]. A critical input to the model is the irrigation accession, IA. This also is incorporated in the dimensionless variable A. Where district-scale flux data is used e.g., surface water diversions, length of channels, this can only be converted to IA through a good knowledge of the irrigation area and how it has changed over the history of the irrigation district. In some cases, agronomic experiments have been conducted to estimate the field water balance and hence water-use efficiency factors. Information of this type may be able to highlight how irrigation accessions have changed over time in response to changed technology, rainfall and availability of surface water. While PerTy3 has not represented this component of the water balance, its accuracy is dependent on good information on surface water balance. The need for drainage is a sign that there is rejected recharge and is useful for the calibration of soil hydraulic parameters [19].

5.5. Further Work

The work above has shown that the processes are well understood; that the 1D modelling of these processes appears to be adequate, but the 2D modelling is only adequate for some aspects. In particular, the assumption that the ponded head is effectively zero external to the irrigated field is clearly inadequate. Some improvement to the model are, therefore, required, perhaps by considering the quasi-steady state analytical models [26], in which a non-zero perched head is incorporated. The 2D transfer function does not vary significantly from the 1D modelling, suggesting that the one transfer function may be adequate for a wide range of situations, and that an exponential model appears to be an adequate approximation. This suggests that such approximants may be useful for estimating recharge, but this requires more work. Brownfield irrigation developments appear to only differ from greenfield developments for large B parameters and for separations of less than 250 m. This suggests that treating developments as independent may be a reasonable approximation of a range of situations. However, more work would be required for the other situations to find the best estimation method.

6. Conclusions

This paper has described the modelling of recharge from new irrigation developments, where there is the potential for perching over a deep vadose zone. The model relies on a surface water balance of the irrigation area and is intended to provide input to regional groundwater models. The model adapts algorithms for which there is no perching to those with perching, allowing a smooth transition through the parameter A.

The work shows that hydrological equilibrium between irrigation accession inputs and recharge to the water table can occur by increased hydraulic gradient through the clay and a greater area of ponded infiltration through the clay. Where the head of the perched water table is sufficiently close to the land surface, some water is returned to the land surface either through sub-surface drainage or by evapotranspiration. There is good agreement between estimates of recharge, ponded head and return to the surface from the semi-analytical PerTy3 model and those from the numerical FEFLOW model for one-dimensional situations, where recharge occurs beneath the irrigated fields. The perching leads to some proportion of the recharge occurring after a long time delay from the timing of the new development.

The model shows that for the parameter range chosen here the effect of two-dimensionality on the total recharge appears to be minor, whereas more detailed processes, such as the proportion of recharge that occurs external to the irrigation field and the ponded head, are sensitive. These latter lateral effects scale in a predictable fashion to a parameter B. Since recharge models usually only require the total recharge, the lack of sensitivity should simplify the modelling.

The modelling also shows that the recharge under brownfield developments, which are developed near pre-existing developments do not appear to be significantly different to that from greenfield developments. This allows the recharge under realistic and complex scenarios of irrigation development to be treated through superposition of recharge from individual fields.

The main input to PerTy3 is the irrigation accession under irrigation. The accuracy of the recharge outputs from PerTy3 is dependent on good information for irrigation accessions over the history of the irrigation development. This, in turn, is dependent on knowing how the irrigation area, and water-use efficiency has changed over time.

Author Contributions: Conceptualization, G.R.W., D.C. and T.S.; methodology, G.R.W., T.S.; software, G.R.W., T.S.; validation, G.R.W. and T.S.; formal analysis, G.R.W.; investigation, G.R.W. and T.S.; resources, D.C., G.R.W.; data curation, G.R.W., T.S. and D.C.; writing—original draft preparation, G.R.W. writing—review and editing, D.C.; project administration, D.C.; funding acquisition, D.C. All authors have read and agreed to the published version of the manuscript.

Funding: This research was partially funded by MURRAY-DARLING BASIN AUTHORITY, project number MD004683.

Acknowledgments: The authors would like to thank the members of the Technical Committee (Juliette Woods, Ray Evans, Emmanuel Xevi and Prathapar) and Hugh Middlemis for technical advice.

Conflicts of Interest: The authors declare no conflict of interest.

References

1. Kurtzman, D.; Scanlon, B.R. Groundwater Recharge through Verti sols: Irrigated Cropland vs. Natural Land, Israel. *Vadose Zone J.* **2011**, *10*, 662–674. [CrossRef]
2. Khan, S. Rethinking rational solutions for irrigation salinity. *Australas. J. Water Resour.* **2005**, *9*, 129–140. [CrossRef]
3. Williams, J.; Grafton, R.Q. Missing in action: Possible effects of water recovery on stream and river flows in the Murray-Darling Basin, Australia. *Australas. J. Water Resour.* **2019**, *23*, 78–87. [CrossRef]
4. Cook, P.G.; Jolly, I.D.; Walker, G.R.; Robinson, N.I. From drainage to recharge to discharge: Some timelags in subsurface hydrology. *Dev. Water Sci.* **2003**, *50*, 319–326. [CrossRef]
5. Jolly, I.D.; Cook, P.G.; Allison, G.B.; Hughes, M.W. Simultaneous water and solute movement through unsaturated soil following an increase in recharge. *J. Hydrol.* **1989**, *111*, 391–396. [CrossRef]
6. Rossman, N.R.; Zlotnik, V.A.; Rowe, C.M.; Szilagyi, J. Vadose zone lag time and potential 21st century climate change effects on spatially distributed groundwater recharge in the semi-arid Nebraska Sand Hills. *J. Hydrol.* **2014**, *519*, 656–669. [CrossRef]
7. Bouwer, H. Effect of Irrigated Agriculture on Groundwater. *J. Irrig. Drain. Eng.* **1987**, *113*, 4–15. [CrossRef]
8. Orr, B.R. *A Transient Numerical Simulation of Perched Groundwater Flow at the Test Reactor Area, Idaho National Engineering and Environmental Laboratory, Idaho, 1952–1994*; U.S. Geological Survey Water Resources Investigations Report 99-4277; U.S. Geological Survey: Denver, CO, USA, 1999.
9. Truex, M.J.; Oostrom, M.; Carroll, K.C.; Chronister, G.B. *Perched-Water Evaluation for the Deep Vadose Zone Beneath the B, BX, and BY Tank Farms Area of the Hanford Site*; Pacific Northwest National Laboratory: Richland, WA, USA, 2013.
10. Oostrom, M.; Truex, M.J.; Carroll, K.C.; Chronister, G.B. Perched-water analysis related to deep vadose zone contaminant transport and impact to groundwater. *J. Hydrol.* **2013**, *505*, 228–239. [CrossRef]
11. Weiss, M.; Gvirtzman, H. Estimating Ground Water Recharge using Flow Models of Perched Karstic Aquifers. *Groundwater* **2007**, *45*, 761–773. [CrossRef] [PubMed]

12. Ganot, Y.; Holtzman, R.; Weisbrod, N.; Nitzan, I.; Katz, Y.; Kurtzman, D. Monitoring and modeling infiltration-recharge dynamics of managed aquifer recharge with desalinated seawater. *Hydrol. Earth Syst. Sci.* **2017**, *21*, 4479–4493. [CrossRef]

13. Rains, M.C.; Fogg, G.E.; Harte, T.; Dahlgren, R.A.; Williamson, R.J. The role of perched aquifers in hydrological connectivity and biogeochemical processes in vernal pool landscapes, Central Valley, California. *J. Hydrol.* **2013**, *505*, 228–239.

14. Villeneuve, S.; Cook, P.G.; Shanafield, M.; Wood, C.; White, N. Groundwater recharge via infiltration through an ephemeral riverbed, central Australia. *J. Arid Environ.* **2015**, *117*, 47–58. [CrossRef]

15. Murray-Darling Basin Authority. Basin Salinity Management 2030 (BSM2030). Murray-Darling Basin, Canberra. 2015. Available online: https://www.mdba.gov.au/publications/mdba-reports/basin-salinity-management-2030 (accessed on 26 March 2020).

16. Besbes, M.; de Marsily, G. From infiltration to recharge: Use of a parametric transfer function. *J. Hydrol.* **1984**, *74*, 271–293. [CrossRef]

17. Mattern, S.; Vanclooster, M. Estimating travel time of recharge water through a deep vadose zone using a transfer function model. *Environ. Fluid Mech.* **2010**, *10*, 121–135. [CrossRef]

18. Wang, X.S.; Ma, M.-G.; Li, X.; Zhao, J.; Dong, P.; Zhou, J. Groundwater response to leakage of surface water through a thick vadose zone in the middle reaches area of Heihe River Basin, in China. *Hydrol. Earth Syst. Sci.* **2010**, *14*, 639–650. [CrossRef]

19. Walker, G.R.; Currie, D.; Smith, A. Modelling the effect of efficiency measures and increased irrigation development on groundwater recharge through a deep vadose zone. *Water.* under review.

20. Currie, D.; Laattoe, T.; Walker, G.R.; Smith, A.; Woods, J.; Bushaway, K. Modelling groundwater returns to streams from irrigation areas with perched water tables. *Water.* under review.

21. Schaap, M.G.; van Genuchten, M.T. A Modified Mualem-van Genuchten Formulation for Improved Description of the Hydraulic Conductivity Near Saturation. *Vadose Zone J.* **2006**, *5*, 27–34. [CrossRef]

22. Brooks, R.H.; Corey, A.T. *Hydraulic Properties of Porous Media*; Colorado State University: Fort Collins, CO, USA, 1964.

23. Meissner, T. Relationship between soil properties of Mallee soils and parameters of two moisture characteristics models. In Proceedings of the SuperSoil 2004: 3rd Australian and New Zealand Soils Conference, Sydney, Australia, 5–9 December 2004; Available online: http://www.regional.org.au/au/pdf/asssi/supersoil2004/2018_meissnert.pdf (accessed on 26 March 2020).

24. Meissner, T. *Estimation of Accession for Loxton and Bookpurnong Irrigation Areas*; Department of Environment and Water: Adelaide, Australia, 2019.

25. Diersch, H.-J. *Finite Element Modeling of Flow, Mass and Heat Transport in Porous and Fractured Media*; Springer: Berlin/Heidelberg, Germany, 2014. [CrossRef]

26. Kacimov, A.R.; Obnosov, Y.V. An exact analytical solution for steady seepage from a perched Aquifer to a low-permeable sublayer: Kirkham-Brock's legacy revisited. *Water Resour. Res.* **2015**, *51*, 3093–3107. [CrossRef]

Article

Modelling the Effect of Efficiency Measures and Increased Irrigation Development on Groundwater Recharge through a Deep Vadose Zone

Glen R. Walker [1],*, Dougal Currie [2] and Tony Smith [2]

[1] Grounded in Water, 2/490 Portrush Rd, St Georges, Adelaide 5064, Australia
[2] CDM Smith, Level 2, 238 Angas Street, Adelaide, SA 5000, Australia
* Correspondence: glen.walker@internode.on.net; Tel.: +61-417668824

Received: 2 March 2020; Accepted: 24 March 2020; Published: 26 March 2020

Abstract: Water use measures are being implemented in irrigation areas to make better use of limited water resources and reduce adverse environmental impacts. A semi-analytical model is developed and tested with a numerical model to estimate changes in timing and magnitude of recharge from such measures in irrigation areas to support management of impacts, especially for areas with deep vadose zones and perched water tables. Low hydraulic conductivity of soil layers will lengthen time delays between actions and changes to recharge in addition to limiting the maximum recharge. Despite variations in detailed processes, the recharge outputs from models are surprisingly similar, irrespective of whether lateral effects are major. Superposition may be used to simplify the modelling of the total change in recharge from successive actions, including the initial development. Further simplification is possible, using an exponential conceptual model to approximate recharge responses to individual actions.

Keywords: hydrology; recharge; soil drainage; water use efficiency; perched water

---

1. Introduction

Globally, irrigation areas are undergoing water use efficiency and infrastructure improvements in order to make better use of limited water and to minimize adverse impacts of irrigation. There are increasing efforts [1] to better account for the resultant changed water balances for surface water supplied irrigation areas and to better understand the downstream impacts on water users and environment. Reduced volumes of irrigation water not used by agriculture will often reduce the returns to streams by groundwater, surface drains or a combination of both [2]. Where saline groundwater is a major pathway, such as on lower reaches of the River Murray in Southeastern Australia, reduced returns may improve future stream salinity and health of riparian zones [3]. Estimation of the timing and quantity of changes in irrigation returns would support the management of salinity in this case; and the accounting and management of downgradient impacts, more generally. This, in turn, requires better estimation of the changes in timing and quantity of groundwater recharge.

The timing of reductions in groundwater recharge under deep vadose zones has been studied in a range of contexts [4–8]. The "kinematic wave" approach [9], which estimates the speed at which changes in pressure move towards the water table, is the simplest. Because of the difficulty of providing soil hydraulic properties over spatial scales relevant for irrigation and environmental management, field studies are required to provide credible estimates for time delays; e.g., [5,8]. A "transfer function" approach [6,10] has been used for situations where there are frequent recharge "events" and it is possible to calibrate the distribution of recharge from water table fluctuations.

Perched water is often associated with irrigation areas. Irrigation efficiency improvements may reduce both the volume of perched water table (possibly to zero) and returns to the land surface from perched water through evapotranspiration and sub-surface drainage. It is expected that perched water will affect the timing of pressure changes from changed irrigation on reaching the underlying water table. In addition to analysing perched water tables for irrigation, e.g., [11,12]; perched water tables have been studied for waste ponds [4,7,13] and floodplain aquifers adjacent to ephemeral streams [14,15]. Steady-state analytical [16,17]; groundwater modelling approaches in which unsaturated zone is treated as an aquitard and numerical unsaturated-saturated zone modelling [4,11,13] have been used to analyse perched water tables, including the contribution to recharge.

There are limitations to the application of existing methods for estimating recharge under perched water tables in deep vadose zones. Numerical methods are difficult to apply to the complex spatial and temporal distribution of irrigation practices, especially given numerical stability problems associated with perched water tables. Transfer functions are yet to be developed for perched water tables. Steady-state analytical methods have yet to be applied to complex irrigation systems. This paper is one of three papers [18,19] that seek to overcome these limitations by developing a semi-analytical model, PerTy3, to estimate the recharge from irrigation areas over deep vadose zones where perching may occur. The model shall try to limit complexity through the adaptation of existing kinematic wave models and scaling principles. This paper contributes to this objective by:

a. adapting the PerTy3 model and associated theory, that had previously been described [18] for application to new irrigation developments, in order to represent water use efficiency measures;
b. testing the theory and the model by comparison with a numerical model; and
c. implementing the model for the whole-of-life irrigation sequence, from development, to represent changes in irrigation practice, including efficiency improvements.

Even if the model is successful, the spatial and temporal complexity of irrigation districts would still mean that a physically based model may still be resource-intensive to implement. This paper, in conjunction with [18,19] aim to address this limitation through a second objective; namely seeking a way to implement transfer functions in order to estimate recharge from irrigation areas over a district-wide scale. The aim of the transfer function aims is to relate the change in irrigation accession at the top of the vadose zone to the recharge to the water table at the base. Transfer functions are widely used for designing and testing electronic and control systems. The process can be simplified if the system is linear and superposition applies. This allows individual actions to be modelled and aggregated to represent an irrigation district; thereby, simplifying the computation. The application of superposition to situations with perched water tables is not straightforward as the presence of perching indicates a nonlinear behaviour, and thresholds between different states. The principle of superposition needs to be tested before its application.

The relationship between inputs and outputs could be conceptually based, rather than physically based and, therefore, be calibrated in a similar way to surface hydrology models. Previous recharge studies using transfer functions [6,10] have calibrated transfer functions on the basis of water table fluctuations. The application to deep vadose zones beneath irrigation districts is likely to differ as groundwater mounds slowly evolve. The calibration of the transfer function is likely to be more difficult and would benefit from knowledge of soil processes. Linear reservoir models have been used previously [6] for recharge. Such a model implies that the output (recharge) is approximated as a linear function of the soil water storage. It is possible that this approximation may be reasonable for perched water table situations. This paper contributes to the second objective by:

a. trialling the principle of superposition of recharge models in response to individual actions over a whole-of-life irrigation sequence; and
b. seeking simple conceptual models that can approximate the recharge distribution for individual actions and in particular the linear reservoir model.

The third overall objective of this paper, in conjunction with [18,19] is to link the developed unsaturated zone model to the surface water balance of irrigation districts, as input, and a regional groundwater model; and developing a process for calibrating this integrated model for application to scenario modelling. This paper contributes to this third objective by testing the calibration of the unsaturated zone model.

2. Theory

This section broadly provides the theory to meet the above objectives:

(1) Section 2.1 describes the conceptual model underlying the theory for objective 1. Sections 2.2 and 2.3 contribute to objective 1(a); i.e., adapting the PerTy3 model and associated theory, which had previously been described [18] for application to new irrigation developments, in order to represent water use efficiency improvements. The theory is developed in two stages: (i) representation of the response of an unsaturated zone to an individual action, which reduces irrigation accession, until the system reaches a new equilibrium (Section 2.2.); and (ii) representation of a response to a series of actions, which reduce irrigation accessions (Section 2.3). The theory adapts the PerTy3 model [18] by adapting the kinematic theory [10].

(2) Sections 2.4 and 2.5 contribute to meeting the second objective; namely, to explore the application of transfer functions to estimate recharge from irrigation areas. More specifically, they contribute to objectives 2(a) and 2(b) by (1) defining transfer functions that can be tested for superposition, given the obvious nonlinearities and (2) describing the theory of the delayed linear reservoir conceptual model.

(3) Section 2.5 contributes to the third objective by providing a theory for the drainage volumes in a manner to allow data on drainage volumes to be used for calibration.

2.1. Conceptual Model for Theory

The conceptual model for this paper is similar to Ref. [18] and is shown in Figure 1. There are three semi-infinite layers of homogeneous soils, in which the second layer is of lower permeability. The left boundary condition is a no flow boundary, i.e., this is a line of symmetry and conditions to the left reflect those on the right. The upper boundary is the base of the agronomic zone, the left side of which underlies irrigated agriculture, and to the right side underlies non-irrigated agriculture. The upper boundary condition is a with a downward water flux, irrigation accession, as determined by agricultural practices, including channel leakage and spillage. For the one-dimensional systems discussed below, the whole upper surface is irrigated. The lower boundary condition is the water table, which is assumed to be constant.

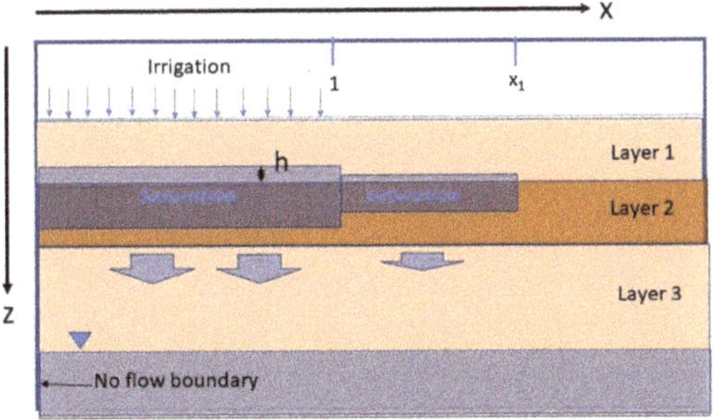

Figure 1. Conceptual model for theory development. The darker brown layer (layer 2) represents the impeding layer. The blue layer represents the saturated zone. The base of layer 1 is the base of the agricultural zone, while the two lighter brown layers represent higher permeability zones. Saturated conditions build up on the base of layer 1 and top of layer 2 (i.e., perching). The irrigation field extends over the horizontal axis, x, from $x = 0$ to $x = 1$. The model dimensions extend beyond the irrigation field (i.e., $x > 1$) to investigate the lateral effects of perching that extend to x_1.

The layers of the unsaturated zone have been parameterised using a modified Mualem–Brooks–Corey model of soil hydraulic properties for each layer [20,21]. The water retention curve is given by:

$$\Theta = \left(\frac{\psi}{hb} \right)^{-\lambda}, \; \psi > h_b \tag{1}$$

$$\Theta = 1, \; \psi \le h_b \tag{2}$$

where ψ is the soil suction, h_b is the air-entry point, λ is a fitting parameter, and the relative saturation, Θ, is given by

$$O = \frac{(\theta - \theta_r)}{(\theta_s - \theta_r)}. \tag{3}$$

The relative permeability, K_r, is given by:

$$K_r = \Theta^m \tag{4}$$

where m is a fitting parameter related to the connectivity of soil pores.

The hydraulic conductivity, K, is obtained by multiplying K_r by the saturated hydraulic conductivity. These parameters will be different for the different layers and a subscript i will be subsequently used to distinguish layer i. The parameters are assumed to be the same for both vertical and horizontal properties, except for the saturated conductivity. A superscript h and v will be used with the saturated hydraulic conductivity to characterise anisotropy. The values given to the parameters is given in the Methods section.

The underlying equations are non-dimensionalised using the vertical length scale, l_2, timescale S_2 l_2/K_{s2}^v, horizontal length scale x_0; where l_i is the thickness of the ith layer; x_0 is the half-width of the irrigated area; K_{si}^v is the saturated vertical conductivity of the ith layer; and S_2 is the specific yield for the second layer for the initial dry conditions. The purpose of the non-dimensionalisation is to simplify as much as possible using scaling and non-dimensional variables.

2.2. Modelling Individual Actions

In this section, the theory for recharge response to a reduction in irrigation accession is developed. The system is assumed to be initially in equilibrium and the change in irrigation accession occurs at time zero. The theory is adapted from the kinematic wave model [10] in which the initial and final states are unsaturated zone and processes are one-dimensional (vertical). This is described in Section 2.2.1. In Section 2.2.2, this theory is adapted to situations, in which the initial state is perched. For perched water tables, water can move laterally from the irrigation district and process can be two-dimensional. Under certain conditions, these two-dimensional processes can be reasonably represented by one-dimensional processes, for which the modelling is simpler.

2.2.1. Kinematic Wave Theory for Unsaturated Zones

Where the original irrigation accession is sufficiently small, the soil zone is initially unsaturated for all layers. In the application of the kinematic wave approach [10] to the above conceptual model, the reduction in irrigation accession at the top of layer one causes the moisture content and soil potential there to reduce to the value for which the intrinsic vertical hydraulic conductivity, K_{r1}, equals the new non-dimensional irrigation accession flux, A. The kinematic value approach follows how any particular value of vertical hydraulic flux (and associated moisture content and soil potential) between the old and new irrigation accession fluxes travels through layer one. The higher value of flux (and associated moisture content and soil potential) travels more quickly, with the fastest being that for the old irrigation accession and the slowest being that for the new irrigation accession. When the fastest value reaches a particular depth, the flux at that depth begins to gradually reduce from the old irrigation accession until the slowest value reaches that depth and flux stabilizes at the new value. Any value of flux can be followed through the three layers to the water table. The theory implies that the speed of any value in the *i*th layer is given by:

$$\frac{dz}{dt} = dK/d\theta = K_{ri}\left(\Theta\right)/\left(\theta - \theta_{ri}\right) G_i \tag{5}$$

where z is the non-dimensional depth of the soil potential below the land surface, and θ is the moisture content, θ; in that layer that corresponds to the soil potential; and:

$$G_i = m_i\, S_i\, K_{si}{}^v/K_{s2}{}^v \tag{6}$$

Recharge will remain at the old value until the fastest value reaches the water table and reduce until the slowest value reaches there and will then stabilize at the new value. The recharge between these values can be obtained by interpolating the times for other values to reach the water table.

Should there be any further measures that reduces irrigation accession, then the above steps can be repeated. Since the group velocity for these new values are lower, there should be no interference between the steps.

2.2.2. Soil Zone Initially with Perched Water

In this section, we consider the situation for which there is a perched water table at the base of the first layer but does not intercept the upper surface of layer 1 and this remains so throughout the transition from old accession rate to new accession rate. It will be shown in this section that most state variables of interest will move exponentially from the old equilibrium value to the new equilibrium value.

If layer one is sufficiently thick, equilibrium conditions can be reached in which the infiltration through the clay equals that of the irrigation accession [18]. The non-dimensional thickness of the perched water above the second layer, h_{eq}, under equilibrium conditions is given by:

$$h_{eq} = (A - 1 - \varphi)/(1 + sqrt(B)) \tag{7}$$

where A is the non-dimensional irrigation accession; B and φ are dimensionless variables, defined by:

$$B = K_{s1}{}^h\, l_2{}^2/(K_{s2}{}^v\, x_0{}^2) \tag{8}$$

$$\varphi = (A - 1) \int d\psi/((A/K_{r2}{}^v\,(\psi) - 1\,)\, l_2) \tag{9}$$

and the integral in Equation (10) is from h_b to ∞ and φ is assumed to be insensitive to A.

Perching only occurs if

$$A > 1 + \varphi \tag{10}$$

and does not intercept the upper boundary condition if

$$l_1/l_2 > (A - 1 - \varphi)/(1 + sqrt(B)) \tag{11}$$

Under conditions of quasi-steady state perched water, and a perched water of zero thickness under non-irrigated agriculture in the fringes of the irrigation areas, the transient condition is given by [18]:

$$\beta\, \frac{\partial h}{\partial t} = A - 1 - \varphi - h\,(1 + sqrt(B)) \tag{12}$$

where β is the ratio of s_1 to S_2; and s_1 is the specific yield of layer 1 after wetting front has already passed through. This has the solution:

$$h = h_{eq} + exp(-t\,(1 + sqrt(B))/\beta)\,(h_0 - h_{eq}) \tag{13}$$

where h_{eq} and h_0 are respectively the new equilibrium head and the old equilibrium head. This shows that perched head reduces exponentially to the new equilibrium conditions. The effect of two-dimensional flow is indicated by the dimensionless parameter B, where $B = 0$ corresponds to one-dimensional flow and large B corresponds to significant lateral movement. The effect of B is to not only reduce the steady-state ponded head, but to also quicken the rate at which the new equilibrium is attained. To understand this better, we consider the dimensioned time scale in the exponential function:

$$t_s = l_2 \, S_2 \, \beta/((1 + sqrt(B)) \, K_{s2}{}^v) \tag{14}$$

As B becomes very large, t_s becomes in dimensioned variables:

$$t_s \sim x_0 \, s_1/sqrt(K_{s1}{}^h \, K_{s2}{}^v) \tag{15}$$

Hence, the time scale involves a mixture of the horizontal conductivity of the sand layer and the vertical conductivity of the clay layer. This is not surprising given that these soil parameters influence both the magnitude of the ponded head and the ability of the water to move laterally. In parallel with the ponded head increasing exponentially to the equilibrium value, the vertical flux through the second layer under the irrigated agriculture would also approach the equilibrium value, q_{eq}:

$$q(t) = q_{eq} - (q_{eq} - q_0) \, exp(-(t - t_0)(1 + sqrt(B))/\beta) \tag{16}$$

where q_0 is the vertical flux at initial equilibrium and is related to h_0 by:

$$q_0 = 1 + h_0 + \varphi \tag{17}$$

There is a similar relationship between q_{eq} and h_{eq}. The extent of the wetting outside of the irrigation field, x_1, also decreases exponentially under the assumptions used from the original equilibrium value, x_{10}, to the new equilibrium value, x_{1eq}, in parallel with the ponded head.

$$x_1(t) = x_{1eq} - (x_{1eq} - x_{10}) \, exp(-(t - t_0)(1 + sqrt(B))/\beta) \tag{18}$$

where x_{10} is related to h_0 by:

$$x_{10} = h_0 \, sqrt(B) + 1 \tag{19}$$

However, the aggregated vertical flux through the clay external to the irrigation field is no longer proportional to the extent of wetting, as areas that had wetted previously will continue to drain. This additional term is ignored in the current modelling.

2.2.3. Change from Perched Water Table to None

If the new irrigation accession is such that $A < 1 + \varphi$, there is no perched water in the new equilibrium situation. The solution to Equation (12) in that situation is still given by Equation (13), but with the term h_{eq} substituted by $(A - 1 - \varphi)/(1 + sqrt(B))$. This allows the time for the perched head to go to zero to be estimated. After that time, layers 2 and 3 begin to drain; and Equations (5) and (6) apply.

2.2.4. Initial State with Rejected Recharge

For the situation where the head of the perched water under the irrigated agriculture intersects the upper boundary (i.e., the first layer is saturated under irrigation), Equation (11) no longer applies and some of the irrigation accession is returned to the surface as there is no opportunity of any further infiltration. We refer to this situation as rejected recharge. If this is the starting situation, then a minor reduction in the irrigation accession may have no effect on recharge, but rather the volume of rejected recharge is reduced. It is only when the irrigation accession is reduced sufficiently for Equation (12) to hold that recharge will change in response to the reduced irrigation accession. The volume of rejected recharge should reduce to zero almost immediately and the head of perched water will also begin to respond immediately. The Equations (12), (13), (16) and (18) should then apply.

2.3. Modelling of Multiple Actions

In Section 2.2.1, the theory for a soil zone with no perched water showed that the response of successive actions, which reduced irrigation accession, were independent of each other. This is unchanged for perched water tables. As can be seen in Equation (12), where it does not matter if the initial condition is an equilibrium situation. The value of the head at the time of the water use efficiency change is used for h_0. Taken together, the methodology across the range of situations described in Sections 2.2.1–2.2.4 does not change under multiple actions, which reduce irrigation accession.

For the first irrigation water use efficiency improvement following a new development, the soil zone may not have come to a hydraulic equilibrium at the time this improvement occurs. Two situations are considered: (1) the wetting front has reached the water table, but perched water has not equilibrated and (2) wetting front is in layer three when new measure occurs.

In the first situation, the effect of reduced irrigation accession should almost be immediate. The perched head should move exponentially to the new equilibrium value or zero (if the new equilibrium state has no perched water). The flux through the second layer should change exponentially in response and the system behaves as in Section 2.2.2. The change in flux will then cause a change in recharge when the pressure effect reaches the water table. Where there is no perched water in the new equilibrium state, layers two and three begin to drain

In situation (2), the perched head will still respond exponentially, as will the flux through layer two. There is some chance that the pressure effect of this may reach the wetting front before it reaches the water table, causing a slight modification in the change in flux at the wetting front and the speed at which it moves. Once the wetting front has reached the water table, the system will respond as per Section 2.2.2.

2.4. Transfer Function and Superposition

Section 2.3 contributes to the second objectives; namely developing a way to implement transfer functions for the estimation of recharge from irrigation areas at district-wide scales. More specifically, it defines modified transfer functions for which superposition may apply. It also develops the theory for the linear reservoir model, in a way that is relevant to irrigation systems with perched water.

A transfer function is a mathematical model of a system that maps its input to its output (or response). A normalised transfer function $TF(t)$ is normally defined as:

$$TF(t) = (R(t) - Ro)/(IA_n - IA_o) \tag{20}$$

An analogous transfer function can be defined for drainage volumes ($D(t)$). These transfer functions will mostly be used for displaying results in Section 4. Where there is no rejected recharge, $TF(t) = 0$ for $t = 0$; and approaches 1 for large times. Hence, it represents a cumulative probability distribution for time delays for pressure to travel through the vadose zone. Where there is rejected recharge, $TF(t)$ is less than one for large times but the sum of transfer functions for recharge and drainage approaches one.

As the system is not necessarily linear, there is no reason to believe that superposition would necessarily apply to a range of situations. Some situations are not expected to follow superposition include those where thresholds are involved, such as the initiation of perching or rejected recharge. The addition of two actions, that do not meet individually exceed the threshold, but in aggregation do so, is not linear. The transfer functions, as defined in Equation (20), is clearly not appropriate for superposition, because of the threshold where rejected recharge occurs. For the purpose of superposition, a modified transfer function for the recharge is defined:

$$TF'(t) = (R(t) - Ro)/(IA_n{}^* - IA_o{}^*) \tag{21}$$

where the superscript * indicates the minimum of irrigation accession (*IA*) and the maximum irrigation accession without rejected recharge. We can define a similar transfer function for drainage, *TF''*, where

$$TF''(t) = (D(t) - D_0)/(IA_n{}^{**} - IA_o{}^{**}) \tag{22}$$

where D_0 is the original drainage rate, IA^{**} is the maximum of *IA* and the maximum irrigation accession that occurs without rejected recharge.

If superposition does apply, the aggregate transfer function for a sequence of actions that affect irrigation accession:

$$TF'(t) = \left(\sum\nolimits_j (IA^*_{j+1} - IA^*_j)\, TF'_{j+1} \right)/(IA^*_{p+1} - IA^*_0) \tag{23}$$

where IA_j is a sequence of modified irrigation accessions that occur from $j = 0$ to $j = p + 1$ and TF'_{j+1} is the modified transfer function that applies for a change of irrigation accession from IA^*_j to IA^*_{j+1}.

The pattern of the aggregate *TF'* with time would broadly follow that of *IA*, but with (a) time delays; (b) "smearing" and (c) capping at the maximum irrigation accession for which there is no rejected recharge. Where superposition applies, this would simplify the numerical process by generating the transfer function for the combination of actions from the transfer function of individual actions.

Within the parameter range used, superposition provided a reasonable approximation to new developments because of the lack of hydraulic interaction between developments [18] i.e., spatial superposition. In this paper, we will test the principle of superposition for a set of actions acting on the one irrigation development at different times, i.e., temporal superposition. For the modelling in this paper, a new development is assumed to be followed by a succession of water use efficiency improvements.

2.5. Theory for the Linear Reservoir Model

The transfer functions for application in Equation (23) can be generated from PerTy3, using the methodology described in Section 2.1. However, one of the objectives of the paper is to explore the application of conceptual models for the transfer function and, in particular, the linear reservoir model. The separation between physically based and conceptual models is not black and white. While the theory above describes physical processes, the analytical model is simplified, with assumptions such as spatial homogeneity, flat surface elevations etc. It is possible to consider transfer functions as conceptual models, using enough complexity to broadly replicate the recharge response. Such models have previously been used for recharge [6,10].

The physical processes above are consistent with a linear reservoir, for which the inputs, *IA*, and outputs, *R*, are related through the storage, *M*:

$$dM/dt = IA^* - q \tag{24}$$

where *M* is the total amount of water in the unsaturated zone, including the perched water table and IA^* is the irrigation accession, adjusted for rejected recharge, if it occurs. If *R* is linear with respect to *M* i.e.,

$$q = cM + d \tag{25}$$

This leads to M being exponential with respect to time with coefficient $-c$. There is a time lag between *q* and *R* due to the time for pressure fronts to move through the unsaturated zone. This leads to:

$$\Delta R = \Delta IA\, (1 - exp(- c\, (t - t_5))),\ t > t_6 \tag{26}$$

where Δ represents the change in the parameter, and c, t_5 and t_6 become fitted parameters. While the theory in previous sections give physical meaning to parameters c and t_l or fitted to model outputs, they can be also fitted to groundwater responses.

The exponential nature of the linear reservoir function is similar to that of many of the parameters in PerTy3. This is perhaps not surprising, given that the mass balance Equation (24) bears striking similarity to Equation (12).

2.6. Criteria for Rejected Recharge

Rejected recharge occurs when Equation (11) is not satisfied. The criterion for rejected recharge can be written in dimensioned form:

$$IA > K_{s2}{}^{v} \left(1 + \varphi + l_1/l_2 + sqrt(K_{s1}{}^{h}/K_{s2}{}^{v})\, l_1/x_0\right) \tag{27}$$

Equation (22) separates soil-related parameters (right-hand side) from irrigation accession (left hand side). This aids the parameterization of the transfer function, as over time, irrigation accession will change and gradually decrease. For many areas, rejected recharge will require drainage of some form for agriculture to be sustained. The presence or absence of drainage and the drainage volume for different irrigation accessions and different soils to calibrate the parameters in the transfer function. Where Equation (23) applies, the drainage volume, D, can be written as:

$$D = IA - K_{s2}{}^{v} \left(1 + \varphi + l_1/l_2 + sqrt(K_{s1}{}^{h}/K_{s2}{}^{v})\, l_1/x_0\right) \tag{28}$$

where data for drainage volumes exist, they can also be used for calibration.

3. Methods

This section describes the methodology to meet the objectives of the paper. Section 3.1 provides a description of the modelling experiments used to address objectives 1 and 2. In particular, the experiments contribute to 1(a) testing the theory and the model by comparison with a numerical model; 1(b) implementing the model for the irrigation whole-of-life; 2(a) trialling the principle of superposition of recharge models in response to individual actions over the irrigation whole-of-life; and 2(b) seeking simple conceptual models that can approximate the recharge distribution for individual actions and in particular the linear reservoir model.

Sections 3.2 and 3.3 describe modelling experiments that contribute to objective 2; namely:

a. trialling the principle of superposition of recharge models in response to individual actions over the irrigation whole-of-life; and
b. seeking simple conceptual models that can approximate the recharge distribution for individual actions and in particular the linear reservoir model.

Sections 3.4 and 3.5 contribute to objective 3 by providing information to support the calibration of integrated models for application to scenario modelling.

3.1. Modelling Experiments

A series of modelling experiments have been conducted to achieve objectives 1 and 2. The models are partitioned into one-dimensional (1D) modelling or two-dimensional (2D) modelling. For the 1D modelling, the variable B has effectively been set to zero. For numerical models, the whole of the upper surface is irrigated and the right-hand boundary has zero lateral flux of water. The degree to which lateral transport is important for the 2D modelling is dependent on the parameter B. This has been set to one for this paper.

Either (or both) the PerTy3 and numerical models have been used. The PerTy3 semi-analytical model has been designed to satisfy the above theory. This does not use a mesh, but rather is designed to estimate time delays. Fluxes are generally estimated at annual time-steps. For transmission of pressure through the unsaturated zone, the time delays are estimated for steps in soil suction and then interpolated to provide recharge at annual time-steps.

The numerical modelling is undertaken using FEFLOWTM [22]. FEFLOW is an acronym of Finite Element subsurface FLOW simulation system and solves the governing flow equations in porous media for variable saturation. Richards' equation is solved for a single dominant fluid phase (in this case water) with an assumed stagnant air phase that is at atmospheric pressure everywhere. FEFLOW implements a number of empirical and spline models for variable saturation and in this work the empirical Mualem–Brooks–Corey model is used [20,21]. Fine mesh refinement is used to achieve stable numerical solutions for the adopted choices of the Mualem–Brooks–Corey model parameters. The set-up for both 1D and 2D modelling is shown in Figure 2. For the 1D modelling (Figure 2a), the third layer has a thickness of 15 m, while for the 2D modelling, it has a thickness of 5 m. The vertical mesh size for both is 10 cm (250 (1D) or 150 (2D) elements), while for the 2D modelling, the horizontal mesh size is 10 m (200 elements).

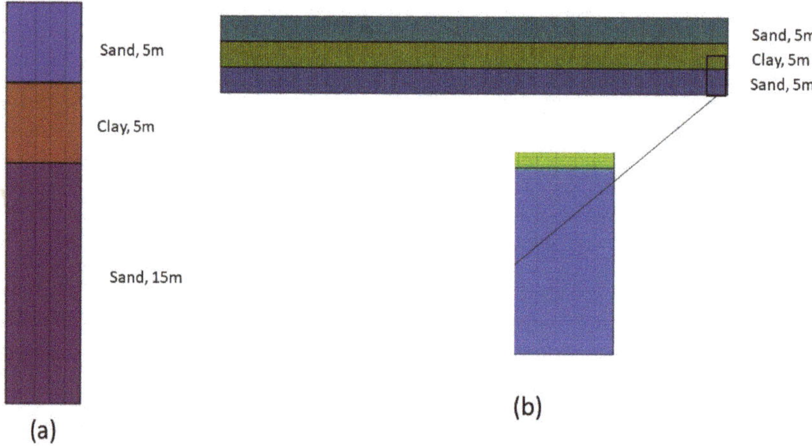

Figure 2. Discretisation for the FEFLOW (Finite Element subsurface FLOW) modelling of (**a**) one-dimensional (1D) situations and (**b**) two-dimensional (2D) situations.

The models have mostly been parameterized using values relevant to the Mallee region of Southeastern Australia [23]. These default values are shown in Table 1. In addition to the parameters in the Tables, the following default values were used:

- irrigation half-width, x_0, 500 m for 2D modelling;
- dimensionless parameter B: 0001 (1D); 1 (2D); and
- pre-irrigation accession for new development: 10 mm/year.

Table 1. Default soil parameters used in the modelling.

Parameter	Symbol	Layer 1	Layer 2	Layer 3
Texture		Sandy Loam	Clay	Sand
Saturated volumetric water content (cm^3/cm^3)	θ_{si}	0.35	0.4	0.38
Residual water content (cm^3/cm^3)	θ_{ri}	0.03	0.1	0.04
Air-entry potential (cm)	h_{bi}	12.0	40.0	8.0
Mualem exponent	m_i	8.24	7	6.94
Vertical saturated hydraulic conductivity (cm/day)	K^v_{si}	300	0.03	500
Anisotropy for saturated conductivity (horizontal/vertical)		~0 (1D) 1 (2D)	~0 (1D) 1 (2D)	~0 (1D) 1 (2D)
Thickness (cm)	l_i	500	500	1500 (1D) 500 (2D)

Table 2 lists the modelling experiments, with non-default parameters and whether the semi-analytical and/or numerical model is used. These modelling experiments can be categorized as follows:

(i)   Experiments 7 (1d) and 8 (1D) model the transient behaviour from an equilibrium state to a new equilibrium state in response to a single reduction in irrigation accession. The modelling outputs of PerTy3 and FEFLOW will be compared. In addition, outputs will be used in Sections 3.2 and 3.3 to test superposition and find approximants for the linear reservoir model.

(ii) Experiments 9 (1D) and 14-1 (2D) model the transient behaviour from an equilibrium state to a new equilibrium state in response to multiple water use efficiency improvements. This includes the behaviour of the perched head and for the 2D modelling recharge under the external to the irrigated agriculture. The modelling outputs of PerTy3 and FEFLOW will be compared. The outputs of the two models will be compared to show differences between 1D and 2D situations. Moreover, the outputs will be used in Sections 3.2 and 3.3 to test superposition, including superposition of approximants.

(iii) Experiments 10 (1D) and 15 (2D) model the transient response from an equilibrium pre-irrigation state to a new equilibrium in response to irrigation development and successive irrigation efficiency improvements. Only PerTy3 is used. The modelling outputs are used to in Section 3.2 to test superposition. The outputs for 10a and 15a come from [18].

(iv) Experiment 11 (1D) models the sensitivity of the recharge over time for the irrigation whole-of-life to $K_{s2}{}^v$. The values of $K_{s2}{}^v$ are chosen so that soil zone goes from no perching to perching for almost the entire modelling period.

Table 2. Parameter values that vary between modelling experiments. 'y' or 'n' indicates whether that model has been used for that experiment.

Model Expt Number	Irrigation Accessions (mm/year)	Non-Dimensional Irrigation Accession	PerTy3	FEFLOW
	IA_n	A		
7 (1D)	230 to 100	2.1 to 0.91	y	y
8 (1D)	100 to 50	0.91 to 0.45	y	y
9 (1D)	230 to150 (0y) to 100 (5y) to 50 (10y)	2.1 to 1.36 to 0.91 to 0.45.	y	y
10a (1D)	10 to 230 (1996)	0.09 to 2.1	[18]	[18]
10b (1D)	10 to 230 (1921) to 150 (2001) to 100 (2006) to 50 (2011)	2.1 to 1.36 to 0.91 to 0.45	y	y
10c (1D)	10 to 230 (1996) to 150 (2001) to 100 (2006) to 50 (2011)	0.09 to 2.1 to 1.36 to 0.91 to 0.45	y	n
10d (1D)	Superposition of 10a and 10b		n	n
11a,b,c,d (1D)	10 to 400 (1996) to 150 (2001) to 100 (2006) to 50 (2011)	0.09 to 3.65 to 1.82 to 0.91 to 0.45	y	n
14-1 (2D)	400 to 200 (0y) to 100 (10y) to 50 (15y)	3.65 to 1.82 to 0.91 to 0.45	y	y
15a (2D)	10 to 400 (1976)	0.09 to 3.65	[18]	[18]
15b (2D)	10 to 400 (1961) to 200 (1981) to 100 (1986) to 50 (1991)	0.09 to 3.65 to 1.82 to 0.91 to 0.45	y	y
15c (2D)	10 to 400 (1976) to 200 (1981) to 100 (1986) to 50 (1991)	0.09 to 3.65 to 1.82 to 0.91 to 0.45	y	n
15d (2D)	Superposition of 15a and 15b			

3.2. Superposition Experiments

There are three superposition experiments:

1.   The first is for a series of water use efficiency improvements, using outputs from experiments 7, 8 and 9. The modelling experiment 9 represents a combination of water use efficiency changes of 230 to 100 mm/year and 100 to 50 mm/year, separated by 5 years. The numerical outputs for experiments 7 and 8, which each represent the individual transitions, but with each starting from an equilibrium state, were superimposed, using Equation (21). This superposition is compared to the semi-analytical and numerical modelling outputs for experiment 9.

2. The second experiment (Experiment 10 (1D)) is for a one-dimensional irrigation whole-of-life i.e., a new development followed by a succession of water use efficiency improvements. The PerTy3 model is implemented for three 1D scenarios: 10a a new development that started in 1996; 10b is a sequence of water use measures separated by 5 years, starting in 2001 (the 1996s state of a new development starting in 1921 approximates an initial equilibrium state); and 10c, the same sequence of water use measures following a new development in 1976. The superposition of 10a and 10b (10d) is compared with 10c.

3. The third experiment (Experiment 15(2D)) is for a two-dimensional irrigation whole-of-life. The PerTy3 model is implemented for three 2D scenarios: 15a, a new development starting in 1976; 15b, a sequence of water use measure separated by 5 years and beginning in 1981 (the 1981 state of a new development in 1961 approximates an initial equilibrium state); and 15c, the same sequence of water use measures following a new development in 1976. The superposition of 15a and 15b (15d) is compared with 10c.

The superposition of spatially separate new developments has been previously tested [18]. Water use efficiency measures in separate developments would be expected to be a weaker test of superposition and is not tested further here.

3.3. Seeking Approximants

Approximants are fitted to the FEFLOW transfer function outputs for modelling experiments 7 (1D) and 8 (1D) using Equation (26).

Moreover, the superposition of the approximants for a series of water use efficiency improvements is tested by comparing to the FEFLOW output for the transfer function for modelling Experiment 14-1.

3.4. Using Drainage Outputs for Calibration

Drainage volumes can be potentially used to calibrate the PerTy3 model. To test this, drainage and soil data are obtained for the Loxton–Bookpurnong irrigation district in the Mallee region.

The soils of the district are categorised into six types. The 3a soils are those for which sub-surface drainage was required, while the 3b soils, none was required. The 3b soils are further divided based on sub-surface characteristics and 3a into two for the different irrigation districts.

Drainage volumes have been measured for the Loxton region [24,25], but not for the Bookpurnong area. For the latter, the presence or absence of sub-surface drainage was noted. The drainage volumes and estimated irrigation accessions [24] were averaged for four periods (1970–1990, 1990–2002, 2002–2006, 2006–2013). These four periods correspond to periods with different estimated water use efficiency factors [24]. The period before 1970 is largely ignored, as a comprehensive drainage scheme was not available before this time. Drainage was only assumed to occur for 3a soils.

Table 3 summarises the soil and drainage characteristics for the six soil types. Areas with no drainage for a period are denoted N/D; areas for which the drainage volume is known; while those where drainage occurs but the volume is unknown are denoted D.

To calibrate PerTy3, Equations (27) and (28) are first applied to create contours of drainage volumes using $K_{s1}{}^h$ and $K_{s2}{}^v$ as variables, keeping other variables constant. By varying $K_{s1}{}^h$ and $K_{s2}{}^v$ are then varied to best fit the drainage data in Table 3. One way to constrain non-uniqueness is to have the same values for all 3a soils and another one for all 3b soils.

Table 3. Soil physical properties and drainage responses for different soils across the Loxton–Bookpurnong Irrigation Districts. ND indicates no drainage and D drainage implemented. Φ is a non-dimensional parameter [18].

District L = Loxton B = Book L/B = both	Soil Type	Layer 1 Thickness l_1 (cm)	Layer 2 Thickness l_2 (cm)	φ	D (mm/year)				
					IA (mm/year) 1920–1970	*IA* (mm/year) 1970–1990	*IA* (mm/year) 1990–2002	*IA* (mm/year) 2002–2006	*IA* (mm/year) 2006–2013
					398	339	317	150	83
L	3a_1	250	350	0.43	D	173	151	ND	ND
B	3a_2	400	600	0.25	D	D	D	ND	ND
L/B	3b_1	500	300	0.51	ND	ND	ND	ND	ND
L	3b_2	1200	200	0.76	ND	ND	ND	ND	ND
L/B	3b_3	400	200	0.76	ND	ND	ND	ND	ND
B	3b_4	500	500	0.30	ND	ND	ND	ND	ND

3.5. Whole-of-Life Modelling

The modelling experiment 11(1D) aims to show the sensitivity of the irrigation whole-of-life recharge to $K_{s2}{}^v$. The PerTy3 model is implemented with the following values for $K_{s2}{}^v$: (a) 0.1 (b) 0.05 (c) 0.03 (d) 0.01 cm/day. The perching conditions show the full range of behaviour over this range of $K_{s2}{}^v$: The irrigation accession is varied from 10 to 400 to 150 to 100 to 50 mm/year, separated by 5 years. This should provide some insight on the interaction between *IA* and $K_{s2}{}^v$ and whether the groundwater response may be able to be used to calibrate soil parameters.

4. Results

4.1. 1D and 2D Modelling

Figure 3 shows the 1D modelling outputs for the transfer function for Experiments 7, 8 and 9. The transfer functions outputs from FEFLOW and PerTy3 models generally match well for both 7 and 8. The worst comparison is for Experiment 8 (non-perching), where the numerical result shows greater "dispersion" of recharge.

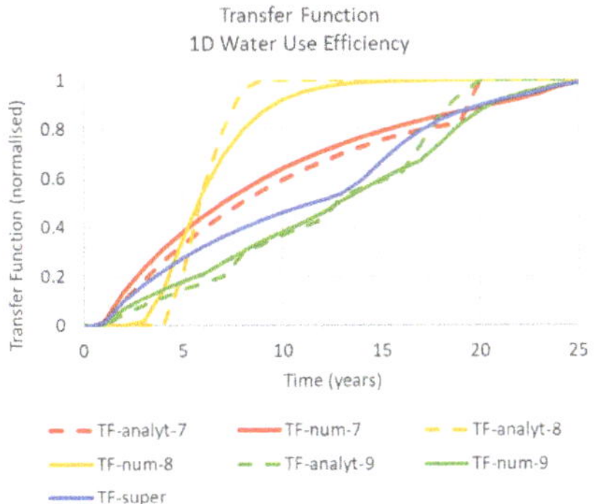

Figure 3. The 1D modelling outputs for Experiments 7 (red), 8 (yellow) and 9 (green). (a) Transfer function (TF). Solid lines indicate FEFLOW (num-erical) outputs, while dashed line shows PerTy3 (semi-analyt-ic) outputs. The superposition for the transfer function (Experiment 9) is denoted by dotted line (super).

Figure 4a shows that for 2D modelling, there is a greater discrepancy between the outputs of FEFLOW and PerTy3. The partitioning between the recharge under irrigated agriculture and external to irrigated agriculture is initially consistent between the two models. However, the recharge for the area external to irrigation is predicted by PerTy3 to decline in about 8 years, compared to 20 years for FEFLOW. Similarly, PerTy3 predicts the recharge under the irrigated agriculture to decline within 15 years compared to 22 years for FEFLOW. The total recharge for both models declines at similar rates to that under the irrigated agriculture.

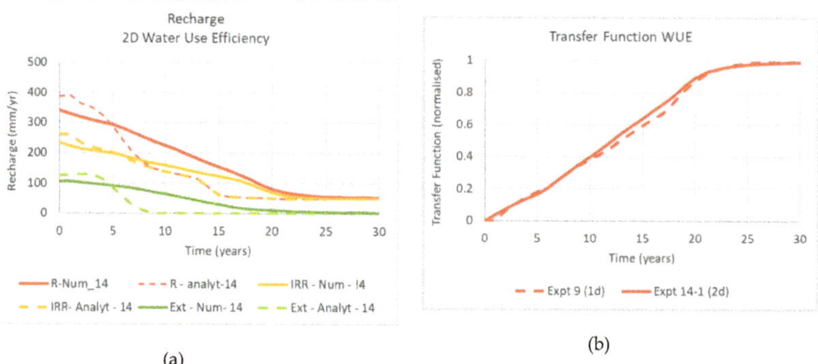

(a)

(b)

Figure 4. (**a**) The 2D modelling outputs for transfer function for experiment 14-1. Solid lines are outputs from FEFLOW (num-erical), while dashed lines are outputs from the PerTy3 (semi-analyt-ic) model. The red lines are for total recharge; yellow that from under irrigated agriculture (IRR) and green for external to irrigated agriculture (Ext). (**b**) Comparison of transfer functions from Experiments 9 (1D) (dashed) and 14-1 (2D) (solid).

Figure 4b shows that the transfer function from the 2D modelling (Experiment 14-1) is almost identical to the that from the 1D modelling (Experiment 9), despite there being a reasonably even split of recharge both under and external to the irrigated agriculture. Moreover, the irrigation accession required for rejected recharge is significantly greater for the 1D situation. There does not appear to be the sensitivity of the exponential time scale as shown in Equation (14).

Figure 5 shows the modelled perched head for both 1D and 2D situations. Again, the outputs are consistent for the 1D situation, while there is a significant discrepancy for the 2D modelling. The perched head is initially consistent, but for the 2D situation, the semi-analytical modelled perched head declines in only 10 years compared to 15 years for the numerical model. This difference might explain that for the recharge in Figure 4. Moreover, the decline in the numerical output is piecewise linear, while that in the semi-analytical model piecewise exponential.

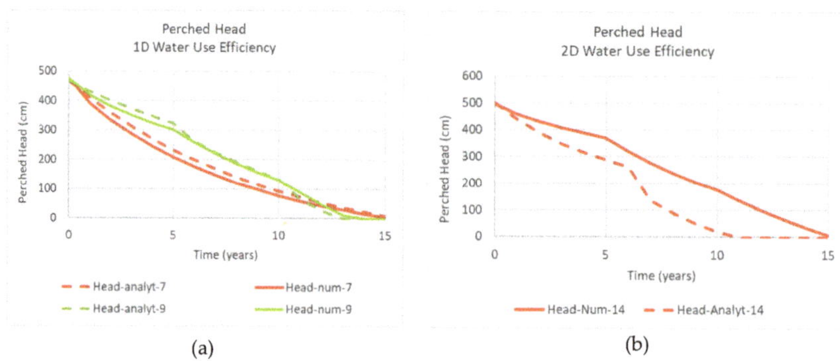

(a)

(b)

Figure 5. Modelled perched head for (**a**) Experiments 7 (1D) (red) and 9 (1D) (green); and (**b**) Experiment 14-1 (2D). Solid lines are FEFLOW (num-erical) outputs, while dashed lines are PerTy3 (semi-analyt-ic) outputs.

4.2. Superposition Experiments

Figure 6 shows the transfer function for the superposition of modelling Experiments 7 and 8 and compares this for both the FEFLOW and PerTy3 model outputs for Experiment 9. While the superposition is higher than either of the outputs for Experiment 9, the shape and time scale of recharge reduction is similar.

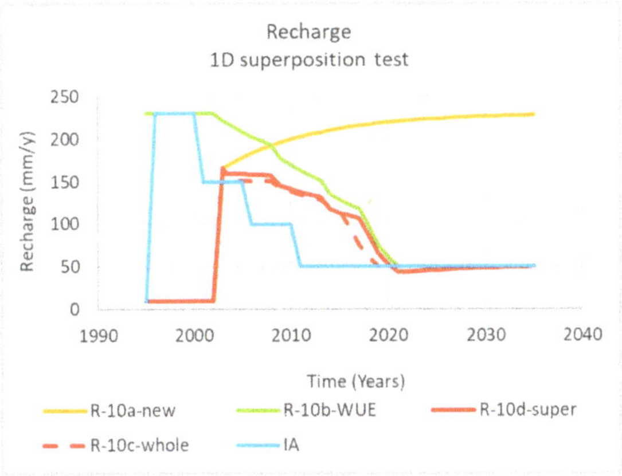

Figure 6. PerTy3 modelling output for Experiment 10 (1D): 10a new development (new, orange line); 10b water use efficiency (WUE, green line); 10c (whole sequence of new development and water use efficiency (whole, dashed red line); Experiment 10d recharge for superposition of 10a and 10c transfer functions (super, red solid line); Irrigation accession for Experiment 10c (1D) blue line.

Figure 6 shows the results of the superposition Experiment 10. It shows the modelled response to the new development and for the subsequent water use efficiency improvements. The superposition (solid red) resembles the modelled output (red dashed). The resultant pattern is a smoothed and delayed version of the irrigation accession (shown in blue). The irrigation accession is the input to the model and reflects initially the pre-irrigation conditions, then the irrigation history from 1976 to 2010 and then the assumed final irrigation accession of 50 mm/year from 2010 onwards.

Figure 7 shows the results of the superposition experiment 15. Again, the superposition (red dashed) matches the modelled recharge (red solid) and the resultant form is a smoothed and delayed version of the irrigation accession (blue). The irrigation accession reflects the irrigation history from pre-irrigation rate of 10 mm/year, development in 1976 and irrigation improvements from 1981 to 1991 and then the final assumed rate of 50 mm/year.

4.3. Use of Drainage Data to Calibrate the Transfer Function

Figure 8a shows contours of drainage volumes (mm/year) for the 3a_1 soil for the Loxton–Bookpurnong District for an irrigation accession rate of 339 mm/year, based on soil hydraulic properties. Negative values mean that there is no drainage. For this value of *IA*, the drainage has been estimated to be 173 mm/year. This implies that the soil properties follow a contour, beginning with $K_{s1}{}^h = 0$ and $K_{s2}{}^v = 0.0212$ cm/day. The drainage volumes for other values of *IA* in Table 3 leads to the same contour. The relationship between the two conductivities for this contour is shown as a solid line in Figure 8b.

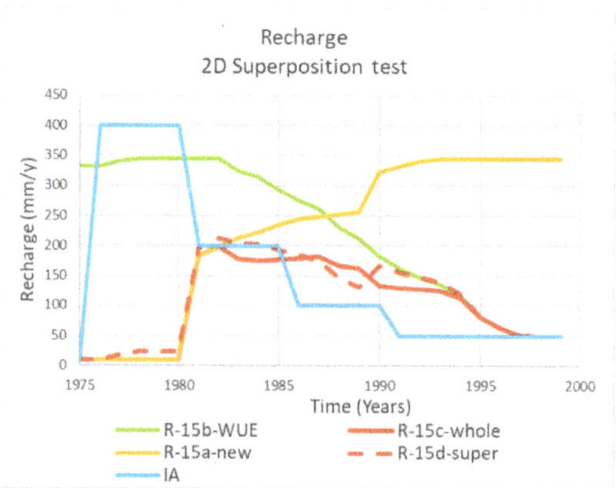

Figure 7. PerTy3 model outputs for Experiment 15 (2D): Experiment 15a new development (new, orange line); Experiment 15b water use efficiency (WUE) measures (green line); Experiment 15c sequence of new development and water use efficiency measures (whole, red solid line); Experiment 15d recharge for superposition of transfer functions (super) for 15a and 15b. The Irrigation accession (IA) is shown in blue.

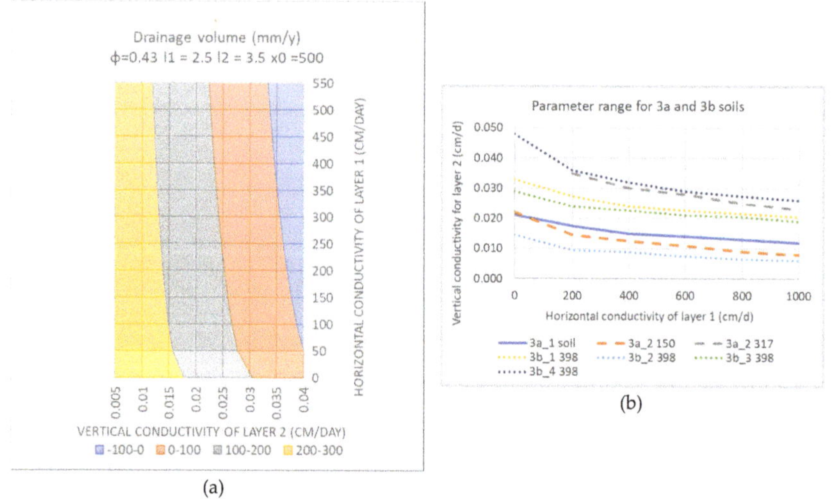

Figure 8. (a) Contours for drainage for the 3a_1 soil type for irrigation accession of 300 mm/year. Negative values correspond to no drainage, while positive values correspond to drainage volumes (mm/year). The resultant drainage for Loxton (173 mm/year) corresponds to the contour starting at the vertical conductivity for layer 2 of 0.0212 and horizontal conductivity of layer 1 of zero. (b) The blue solid line shows this same contour (3a_1 soil). Contours (dashed lines) are also shown for 3a_2 soil for *IA* of 150 (brown, 3a_2 150) for which there is no drainage and 317 (grey, 3a_2 317)) for which there is drainage. This means that the soil properties should lie between these contours and is consistent with the blue contour for all but the lowest horizontal conductivity. Further contours (dotted lines) are shown for 3b soils (3b_1 398; 3b_2 398; 3b_3 398; 3b_4 398). Even for *IA* of 398 mm/year, there is no drainage. Soil properties therefore should lie above these contours. The contour for 3b_4 forms the strongest constraint.

While the presence or absence of drainage is known for soil 3a_2, the drainage volumes are not. This knowledge provides some constraints on the soil properties. The absence of drainage for a value of *IA* of 150 mm/year, but drainage at 317 mm/year means that the soil properties lie between the two dashed contours in Figure 8b. If we are seeking one set of soil properties denoted as 3a, the contour found for 3a_1 would be consistent with this constraint for all but the lowest value of horizontal conductivity.

For 3b soils, there is no drainage, even for a value of *IA* of 398 mm/year. This means that the soil properties must lie above the dotted contours in Figure 8b. If we are seeking a single set of soil properties for all 3b soils, the strongest constraint is given by 3b_4. This implies that $K_{s2}{}^{v} > 0.025$ cm/day for high values of $K_{s1}{}^{h}$ or >0.045 cm/day (low values of $K_{s1}{}^{h}$). Soil properties can then only be further constrained by fitting to the groundwater response.

4.4. Approximants

Figure 9a shows fitted approximants for the transfer functions for FEFLOW outputs for experiments 7 and 8 (1D). The exponential functions with a time delay were reasonable approximations for both the perched situation and the unperched situation. For the latter, the approximant underestimated the later stages.

Figure 9. (**a**) Fitted approximants for 1D FEFLOW (num-erical) transfer functions for modelling experiments (7) and (8). The approximants are respectively $1 - \exp(-0.11 \times (t - 0.8))$ for t > 2; $1 - \exp(-0.32 \times (t - 3.5))$ for t > 4. (**b**) Superposition of a succession of fitted approximants (a) to reductions of irrigation accession from 350 to 200, 200 to 150 (5 years later), 150 to 100 (5 years later) and 100 to 50 (5 years later) and compared to FEFLOW (Num-erical) output for 2D modelling Experiment 14-1.

Figure 9b showed the superposition of these approximants for a series of successive reductions of irrigation accession from 350 to 200 to 150 to 100 to 50 mm/year at time intervals of 5 years. The superposition of the approximants overestimated recharge during the later stage, but still provided a reasonable approximation for application to groundwater modelling.

4.5. Whole-of-Life Modelling

Figure 10 shows the recharge response to a whole-of-lifetime sequence for irrigation areas. The results show that time delays for new developments increase for decreasing $K_{s2}{}^{v}$. The highest value of $K_{s2}{}^{v}$ (01.1 mm/day) produces the earliest and highest peak value, equalling *IA* after five years. For other values of $K_{s2}{}^{v}$, the peak values occur later and have lower values, with the lowest value of $K_{s2}{}^{v}$ (0.01 mm/day) having a barely discernible peak. For $K_{s2}{}^{v} = 0.1$ mm/day, the shape mirror that of *IA*, but with a delay varying from 3 to 8 years. For the next highest value of $K_{s2}{}^{v}$, 0.05 mm/day, recharge reduces exponentially until there is no perched head and then falls to match the recharge for 0.1 mm/day. For the third highest value of $K_{s2}{}^{v}$, 0.03 mm/day, recharge also falls, more or less, exponentially (but more slowly than for $K_{s2}{}^{v}$ = 0.05 mm/day), to where the perched water disappears and then falls to the final value only two years after the previous curves, and with a time delay of up to twelve years after *IA*. Reduction of $K_{s2}{}^{v}$ and the presence of perched water leads to lower peak recharge values, greater time delays. When perched water disappears, the difference between the recharge disappears.

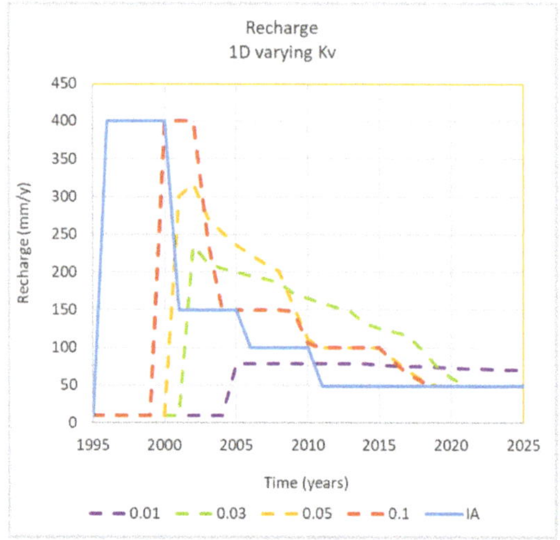

Figure 10. Plots of recharge with time for irrigation systems with different values of $K_{s2}{}^{v}$ (a) 0.01 (b) 0.03 (c) 0.05 and (d) 0.1 cm/day in response to a sequence of *IA* shown as a solid blue line.

5. Discussion

This paper has built upon a semi-analytical model, PerTy3, to estimate the impact of reduced irrigation accessions on recharge, adapting an analytical model [10]. This has focused on deep vadose zones, where time delays for pressure transmission can be significant and where perching can occur. The theory allows straightforward changes in the status of perching as irrigation accession increases after development and reduces after infrastructure and water use measures. In addition, the theory has been generalized to two-dimensional situations. The use of non-dimensional parameters *A* and *B* reflect these transitions, *A* for perching and *B* for lateral flow. Perching will occur where *A* is greater than $1 + \varphi$ and rejected recharge occurs where A is large enough for criterion (Equation (11)) to no longer hold. One-dimensional behaviour will occur when *B* is small and lateral movement becomes significant as *B* becomes greater than one.

5.1. Accuracy of the Model

The model is conceptual in nature and simplifies the processes under irrigation areas. For example, topography is assumed to be flat and sub-surface soil layers are homogeneous in nature. The expectation is that some parameters will need to be calibrated, but the data to do this is limited. The aim is to capture the key processes, while keeping the model as simple as possible. Parameters related to the key processes can then be calibrated.

The testing of the model can occur in three ways:

(1)   model-to-model comparison;
(2)   comparison of assumptions with past literature; and
(3)   testing with field data.

Outputs from the semi-analytical and numerical models have been compared (Figures 3–5). The underlying assumptions and parameters for the models are very similar. The numerical model may not necessarily be taken as a point of truth, as it may have problems with numerical instability and dispersion, especially with the fine mesh required for modelling perched water situations. From inspection, there did not appear to be issues of instability. Closer examination of results than those shown here is possible and has been conducted. The model in this paper used a more restricted set of parameters than in [18] for the new developments and so any findings needs to be qualified that they are for the range of parameters used.

The outputs for PerTy3 and FEFLOW are generally very consistent for one-dimensional situations. Even the closer inspection of outputs shows great consistency. The largest discrepancy appears to be the modelling of the drainage of unsaturated zone, where the semi-analytical model predicted a quicker response than the numerical model. However, this discrepancy occurs for drier conditions, where perching is not an issue and, hence, associated with the analytical model [10]. The most likely issue is that this kinematic wave model only considers the gravity component of Richard's Equation and ignores the diffusive component. The latter becomes more important for drier conditions [26]. The diffusive component would lead to upward flow from the capillary fringe; thus, offsetting the gravity term and slowing the pressure fronts. This would lead to PerTy underestimating time lags for drier conditions.

The discrepancy between outputs from PerTy3 and FEFLOW is much greater for the two-dimensional modelling. The numerical modelling of total recharge is similar to the one-dimensional modelling [18], even for a value of B of 10. Figure 4b shows a similar result for reduced irrigation accessions, although the testing was limited to where B was one. The degree of similarity appears to be surprising, given the detailed processes are different. Other indicators (irrigation accession at which rejected recharge occurs and partitioning between the recharge under the irrigated agriculture are showing significant differences from 1D processes. This similarity may be useful for groundwater modelling, where the total recharge is the quantity of interest.

PerTy3 outputs shows agreement with the numerical modelling in the partition proportions and the initial response for the perched head. However, it is predicting smaller time delays overall for recharge and perched head to reduce. A possible reason for the discrepancy is the assumption that the perched head is not affecting soil infiltration external to the irrigated field. This assumption differs from some of the literature for steady-state solutions e.g., [17]. The perched head for x slightly greater than 1 should be similar to that slightly less than 1. A fall in irrigation accession would lead to not only a change in wetted area, but also a fall in the ponded head external to irrigated agriculture. Consideration of the perched head external to the irrigated agriculture would lead to behaviour closer to the one-dimensional situation.

The large time delays and dispersion as predicted are generally consistent with model inversions to infer recharge from the groundwater response [3]. This is investigated more objectively in Ref. [19].

5.2. Sensitivity and Calibration

The paper has shown how drainage volumes and the absence or presence of drainage may be used to constrain soil properties. Drainage is very sensitive to $K_{s2}{}^v$, but much less so to $K_{s1}{}^h$. The equilibrium perched head is also very sensitive to $K_{s2}{}^v$, but $K_{s1}{}^h$ can significantly reduce its value. It is possible that maps of perched head could be used to constrain both model parameters. The large variation in recharge patterns in Figure 10 for reducing $K_{s2}{}^v$ suggests that groundwater responses could be used to calibrate $K_{s2}{}^v$. Because of the spatial variability of $K_{s2}{}^v$ and the difficulty in measuring $K_{s2}{}^v$ in the field, it is difficult to provide an independent value of $K_{s2}{}^v$ within an order of magnitude, yet the impact on recharge can be very dramatic.

5.3. Superposition and Approximants

This paper describes superposition experiments in time. The transfer function for a new development in the vicinity of an existing development will be approximately the same as that far away from any existing developments for the parameters tested [18]. This paper extends on that work by showing that the change in recharge from a superposition of independent actions is almost identical to that of the same set of actions occurring successively, within the range tested. This means that it is feasible to consider the numerous actions occurring both spatially and temporally across the irrigation district could be considered as independent actions and, hence, the total recharge could be considered as a sum of the changes in recharge from the individual actions. If this is shown to be the case through further testing, this could simplify the analysis of a complex irrigation district. Because there are thresholds with perched water tables and rejected recharge, there are limitations to the range of situations that superposition applies.

In parallel to considering superposition, it may be possible to use simpler approximants for the transfer function. Figure 9a showed that simple exponential functions with a built-in time delay provided a reasonable approximation to 1D transfer functions from numerical modelling. Figure 9b showed that the superposition of these approximants provided a reasonable approximation to modelling experiment 14-1. The delayed exponential function is consistent with linear reservoir modelling used by others for the vadose zone [6]. The modelling shows that there is some potential to use simple additions of transfer functions for individual actions. The fitted exponential decay parameter is strongly related to the vertical conductivity of the impeding layer.

5.4. Learnings from the Modelling

The modelling outputs has shown that the effect of a low conductivity layer is to lengthen time delays between changes in irrigation accession and recharge, perhaps up to fifteen years for the parameters considered here. This has implications for salinity management, as this implies that groundwater pumping or other engineering works may need to be maintained until the impacts of the water use efficiency measures are effective. Where groundwater is fresh, the delays in reduction in irrigation returns to the river from water use efficiency measures may be large. Where decommissioning is being implemented, the time delays are expected to be much longer again.

In summary, there has been significant progress towards efficient modelling of recharge from irrigation areas, especially where the vadose zone is deep and where perched water tables exist. However, the following issues would benefit from more work:

1. The 2D modelling for both new developments and water use efficiency measures: while the current modelling does capture some aspects consistently with the numerical modelling, it underestimates the overall time delays between a change in IA and a change in recharge.
2. Drainage from unsaturated soils: The current model underestimates the time for drainage to occur from an unsaturated soil.
3. Testing of superposition and approximants: The work so far shows that superposition and simple approximants has promise in simplifying the modelling of recharge from irrigation areas. However, it does need to be tested across a broader range of parameters and situations before regular use.

6. Conclusions

This paper describes the development of a model of recharge from an irrigation area undergoing water use efficiency changes in parallel with new development. It specifically addresses irrigation areas overlying perched water tables and deep vadose zones. This is particularly relevant to the Southwestern part of the Murray–Darling Basin in Southeastern Australia, where groundwater mounds under irrigation areas are pushing saline water into the River Murray. The impact on river salinity is being managed through a combination of groundwater interception schemes and water use efficiency measures.

The outputs from semi-analytical model, PerTy3, which has been adapted from an existing model [10], and compared to those from a numerical FEFLOW model. For the one-dimensional (1D) situation, (i.e., modelling of large irrigation areas with low horizontal hydraulic conductivity), the outputs are consistent for situations with perched water tables. Where there is rejected recharge, the recharge will not change until the irrigation accessions reduces so that there is no longer rejected recharge. In the absence of rejected recharge, the perched head (and recharge) falls exponentially in response to reduced irrigation accessions. If the reduction in irrigation accession is such that perched water tables disappear, or there were no perched water tables initially, the model is underestimating the time delays for pressure changes to move through the unsaturated zone as the soil drains.

PerTy3 also appears to underestimating time delays for two-dimensional situations. In these situations, perched water moves laterally across the impeding layer and then infiltrates. As the irrigation accession is reduced, the ponded head falls and area of wetting external to the irrigation field reduces. The 2D modelling predictions of changes in recharge from the numerical model are almost identical to those from the 1D modelling for the parameter range tested here.

The superposition of changes in recharge due to independent actions is close to the change in recharge from a succession of those actions, for the range of testing. If this applies more broadly, it may simplify the modelling of a complex irrigation area, where developments and irrigation water use efficiency measures are occurring over time and space. It also appears that simple approximations may be able to be used, which would further simplify such an analysis.

The main impact of a low conductivity layer is to delay the impact of water use efficiency changes on recharge, due to the time delays for pressure responses to reduced irrigation accession to travel through the vadose zone. This has significance for management, as the delays may mean that interim measures, such as pumping for salinity, may need to continue for longer.

Author Contributions: Conceptualization, G.R.W., D.C. and T.S.; methodology, G.R.W., T.S.; software, G.R.W., T.S.; validation, G.R.W. and T.S.; formal analysis, G.R.W.; investigation, G.R.W. and T.S.; resources, D.C., G.R.W.; data curation, G.R.W., T.S. and D.C.; writing—original draft preparation, G.R.W. writing—review and editing, D.C.; project administration, D.C.; funding acquisition, D.C. All authors have read and agreed to the published version of the manuscript.

Funding: This research was partially funded by MURRAY-DARLING BASIN AUTHORITY, project number MD004683.

Acknowledgments: The authors would like to thank the members of the Technical Committee (Juliette Woods, Ray Evans, Emmanuel Xevi and Prathapar) and Hugh Middlemis for technical advice.

References

1. Perry, C. Efficient irrigation; inefficient communication; flawed recommendations. *Irrig. Drain.* **2007**, *56*, 367–378. [CrossRef]

2. Wang, Q.J.; Walker, G.; Horne, A. *Potential Impacts of Groundwater Sustainable Diversion Limits and Irrigation Efficiency Projects on River Flow Volume under the Murray-Darling Basin Plan: An Independent Review*; Report written for the MDBA; University of Melbourne: Parkville, Australia, 2018; p. 73. Available online: https://www.mdba.gov.au/sites/default/files/pubs/Impacts-groundwater-andefficiency-programs-on-flows-October-2018.pdf (accessed on 26 March 2020).

3. Yan, W.; Li, C.; Woods, J. *Loxton-Bookpurnong Numerical Groundwater Model*; Volume 1: Report and Figures; Technical Report DFW 2011/22; Department for Water: Adelaide, Australia, 2011. Available online: https://www.waterconnect.sa.gov.au/Content/Publications/DEW/DFW_TR_2011_22_Vol_1.pdf (accessed on 26 March 2020).

4. Orr, B.R. *A transient numerical simulation of perched ground-water flow at the test reactor*; US Geological Survey Water-Resources Investigations Report 99–4277; Idaho National Engineering and Environmental Laboratory: Idaho Falls, ID, USA, 1999.

5.   Mattern, S.; Vanclooster, M. Estimating travel time of recharge water through a deep vadose zone using a transfer function model. *Environ. Fluid Mech.* **2010**, *10*, 121–135. [CrossRef]

6.   Wang, X.S.; Ma, M.-G.; Li, X.; Zhao, J.; Dong, P.; Zhou, J. Groundwater response to leakage of surface water through a thick vadose zone in the middle reaches area of Heihe River Basin, in China. *Hydrol. Earth Syst. Sci.* **2010**, *14*, 639–650. [CrossRef]

7.   Truex, M.J.; Oostrom, M.; Carroll, K.C.; Chronister, G.B. *Perched-Water Evaluation for the Deep Vadose Zone Beneath the B, BX, and BY Tank Farms Area of the Hanford Site*; Pacific Northwest National Laboratory: Richland, WA, USA, 2013.

8.   Rossman, N.R.; Zlotnik, V.A.; Rowe, C.M.; Szilagyi, J. Vadose zone lag time and potential 21st century climate change effects on spatially distributed groundwater recharge in the semi-arid Nebraska Sand Hills. *J. Hydrol.* **2014**, *519*, 656–669. [CrossRef]

9.   Sisson, J.B.; Ferguson, A.H.; van Genuchten, M.T. Simple method for predicting drainage from field plots. *Soil Sci. Soc. Am. J.* **1980**, *44*, 1147–1152. [CrossRef]

10.   Besbes, M.; De Marsily, G. From infiltration to recharge: Use of a parametric transfer function. *J. Hydrol.* **1984**, *74*, 271–293. [CrossRef]

11.   Schneider, A.D.; Luthin, J.N. Simulation of Groundwater mound perching in Layered Media. *Trans. ASAE* **1978**, *21*, 920–930. [CrossRef]

12.   Bouwer, H.M. Effect of Irrigated Agriculture on Groundwater. *ASCE J. Irr. Drain. Eng.* **1987**, *113*, 4–15. [CrossRef]

13.   McCray, J.E.; Nieber, J.; Poeter, E.P. Groundwater Mounding in the Vadose Zone from On-Site Wastewater Systems: Analytical and Numerical Tools. *J. Hydrol. Eng.* **2008**, *13*, 710–719. [CrossRef]

14.   Reid, M.E.; Dreiss, S.J. Modeling the effects of unsaturated, stratified sediments on groundwater recharge from intermittent streams. *J. Hydrol.* **1990**, *114*, 149–174. [CrossRef]

15.   Villeneuve, S.; Cook, P.G.; Shanafield, M.; Wood, C.; White, N. Groundwater recharge via infiltration through an ephemeral riverbed, Central Australia. *J. Arid Environ.* **2015**, *117*, 47–58. [CrossRef]

16.   Khan, M.Y.; Kirkham, D.; Handy, R.L. Shapes of steady state perched groundwater mounds. *Water Resour. Res.* **1976**, *12*, 429–436. [CrossRef]

17.   Kacimov, A.R.; Obnosov, Y.V. An exact analytical solution for steady seepage from a perched Aquifer to a low-permeable sublayer: Kirkham-Brock's legacy revisited. *Water Resour. Res.* **2015**, *51*, 3093–3107. [CrossRef]

18.   Walker, G.R.; Currie, D.; Smith, T. Modelling recharge from irrigation developments with a perched water table and deep unsaturated zone. *Water* **2020**, under review.

19.   Currie, D.; Laatoe, T.; Walker, G.R.; Smith, A.; Woods, J.; Bushaway, K. Modelling groundwater returns to streams from irrigation areas with perched water tables. *Water* **2020**, under review.

20.   Schaap, M.G.; van Genuchten, M.T. A Modified Mualem–van Genuchten Formulation for Improved Description of the Hydraulic Conductivity Near Saturation. *Vadose Zone J.* **2006**, *5*, 27–34. [CrossRef]

21.   Brooks, R.H.; Corey, A.T. *Hydraulic Properties of Porous Media*; Hydrology Papers; Colorado State University: Fort Collins, CO, USA, 1964.

22.   Diersch, H.-J. *Finite Element Modeling of Flow, Mass and Heat Transport in Porous and Fractured Media*; Springer: Berlin/Heidelberg, Germany, 2014.

23.   Meissner, T. Relationship between soil properties of Mallee soils and parameters of two moisture characteristics models. In Proceedings of the 3rd Australian and New Zealand Soils Conference, Sydney, Australia, 5–9 December 2004. Available online: www.regional.org.au/au/pdf/asssi/supersoil2004/2018_meissnert.pdff (accessed on 26 March 2020).

24.   Meissner, T. *Estimation of Accession for Loxton and Bookpurnong Irrigation Areas*; Department of Environment and Water: Adelaide, Australia, 2019.

25.   Rolls, J. *Berri, Loxton and Renmark Irrigation Area CDS Water: Implications of Flow and Quality Data*; DWLBC Report 2007/24; Government of South Australia: Adelaide, Australia, 2007.

26.   Pastore, N.; Cherubini, C.; Giasi, C. Kinematic diffusion approach to describe recharge phenomena in unsaturated fractured chalk. *J. Hydrol. Hydromech.* **2017**, *65*, 287–296. [CrossRef]

Article

Modelling Groundwater Returns to Streams From Irrigation Areas with Perched Water Tables

Dougal Currie [1,*], Tariq Laattoe [2], Glen Walker [3,*], Juliette Woods [2], Tony Smith [1] and Kittiya Bushaway [2]

[1]   CDM Smith, Level 2, 238 Angas St., Adelaide, SA 5000, Australia
[2]   South Australian Department for Environment and Water, 81–95 Waymouth Street, Adelaide, SA 5000, Australia
[3]   Grounded in Water, 2/490 Portrush Rd., St. Georges, Adelaide, SA 5064, Australia
*   Correspondence: curriedr@cdmsmith.com (D.C.); glen.walker@internode.on.net (G.W.)

Received: 2 March 2020; Accepted: 25 March 2020; Published: 27 March 2020

Abstract: Quantifying the magnitude and timing of groundwater returns to streams from irrigation is important for the management of natural resources in irrigation districts where the quantity or quality of surface water can be affected. Deep vadose zones and perched water tables can complicate the modelling of these fluxes, and model outputs may be biased if these factors are misrepresented or ignored. This study was undertaken in the Murray Basin in southern Australia to develop and test an integrated modelling method that links irrigation activity to surface water impacts by accounting for all key hydrological processes, including perching and vadose zone transmission. The method incorporates an agronomic water balance to simulate root zone processes, semi-analytical transfer functions to simulate the deeper vadose zone, and an existing numerical groundwater model to simulate irrigation returns to the Murray River and inform the management of river salinity. The integrated modelling can be calibrated by various means, depending on context, and has been shown to be beneficial for management purposes without introducing an unnecessary level of complexity to traditional modelling workflows. Its applicability to other irrigation settings is discussed.

Keywords: hydrology; vadose zone modelling; perched water tables; irrigation recharge; groundwater returns to streams; groundwater modelling

---

1. Introduction

Irrigation often results in return flows to streams, either by drainage systems established to remove excess irrigation water in the root zone (drainage returns) or by excess irrigation water moving past the root zone, recharging groundwater and then discharging to streams (groundwater returns) [1]. Return flows can be significant in terms of quantity and/or quality and can potentially impact downstream users or aquatic and riparian ecology [2]. Quantifying the magnitude, timing, and quality of return flows to streams is required to inform the management of these risks [3,4]. Drainage returns can be measured directly, but modelling is required to quantify groundwater returns [1].

Several processes need to be represented in the modelling in order to estimate the timing and magnitude of groundwater returns [5,6]. Surface and rootzone processes control the amount of water that moves past the rootzone (rootzone drainage) [7]. This water must pass through the remainder of the vadose zone before it recharges the groundwater system (irrigation recharge). Saturated zone processes, including groundwater–surface water dynamics, control the transmission of this flux to streams.

The surface and rootzone processes are comparatively well understood, being an area of focus for agronomic specialists in the interests of devising improved irrigation practices. The saturated zone also receives attention, being an area of focus for hydrologists and hydrogeologists in the interests of water resource management. The deeper vadose zone, being harder to measure and falling between scientific disciplines, is not as well understood.

Where the vadose zone is thin and relatively uniform with depth, there is minimal delay between rootzone drainage and irrigation recharge such that it is reasonable to assume their equivalence. However, where the vadose zone is thick or includes lithological units where perched water tables may form, the timing and magnitude of irrigation recharge can be significantly affected. A thick vadose zone means that it can take a long period of time (in the order of years or decades) for changes in root zone drainage (e.g., due to irrigation development or the introduction of efficiency measures) to propagate to the water table [8–10]. If a layer of relatively low permeability (e.g., a clay unit) occurs within the vadose zone, a perched water table may form above this layer. In these circumstances, the rate of downwards flux will be restricted by the vertical hydraulic conductivity of the low permeability unit and the perched water table may grow vertically and laterally, and some water could be returned to the surface (e.g., via evapotranspiration or via drainage systems) [11–15].

The inclusion of the deeper vadose zone within catchment-scale models is rare, particularly under irrigated settings. Existing vadose zone models, which can be applied at large scales, mostly assume that water moves under gravity (e.g., [16,17]), but this assumption is invalid where perched water tables form [18].

The challenge of including the deeper vadose zone within models has resulted in different approaches to modelling groundwater returns: forward and inverse modelling. The forward method (e.g., [19,20]), aims to model the relationship between irrigation recharge and groundwater returns. The inverse method (e.g., [15,21]) estimates irrigation recharge using observed groundwater level responses or spring discharges and applies the derived irrigation recharge rates to estimate groundwater returns. There are limitations to both methods. The forward method will often involve simplifying assumptions such as ignoring any time lags or losses associated with perching and use uncertain parameters. Errors due to simplifying assumptions and uncertainty are directly transferable to the predicted groundwater returns. The estimation of recharge by inverse method, which is mostly used where there is no appropriate vadose zone model, does not provide unique solutions. It is also difficult to formulate scenarios to model different irrigation practices because on-ground actions and their influence on rootzone drainage are not directly linked to irrigation recharge and groundwater returns within the modelling approach.

Both the forward and inverse methods may lead to biases in estimating groundwater returns from irrigation and create uncertainty when linking on-ground actions (e.g., irrigation efficiency improvements) to impacts on the river. Biases can result in several disbenefits for the management of groundwater returns. If the risks of groundwater returns are overstated or thought to occur more acutely than in reality, control measures might be overdesigned, too expensive, or place an unnecessarily restrictive burden on irrigation development. If the risks are understated or thought to occur more gradually than in reality, control measures may be insufficient to avoid unacceptable impacts. Furthermore, if the modelling approach fails to appropriately link on-ground actions with their effects, there can be an inequitable sharing of the cost burden to manage risks.

Recognising the limitations of current methods, the authors of [22,23] developed a semi-analytical approach to model vadose zone transfers. These transfer functions include the simulation of perching and allow for the time lags inherent in unsaturated flow.

The objective of this study was to make use of the transfer functions developed by [22,23] and develop an integrated modelling method to quantify the groundwater returns from irrigation for the management of salinity in the lower Murray River, Australia. The integrated approach links the various elements of hydrological system involved: (1) the surface/rootzone, (2) the deeper vadose zone, and (3) the saturated zone. The integration of all three elements is intended to reduce the risk of biases that are inherent in current methods by allowing the models to make full use of the datasets available, as well as to create a transparent process to simulate the impact of various management actions. The method must be applicable on a catchment-wide scale and be practical to implement; that is, it should not impose an unnecessary burden on existing model workflows and their data requirements.

2. Study Area Description

2.1. Salinity Management in the Murray River

The Murray River is Australia's longest river. Its catchment (including that of the Darling River) spans more than 1 million km^2. It provides water to more than 1.5 million households, including the city of Adelaide, and supports a substantial irrigation industry, being Australia's primary food bowl. Along the lower reaches of the Murray River, low salinity river water (in the order of 100–200 mg/L) is used to irrigate a range of horticultural crops across several irrigation districts. Irrigation has significantly altered the hydrology of this semi-arid environment, leading to much higher recharge rates under areas of irrigation. Because the groundwater is naturally highly saline (typically 18,000 to 35,000 mg/L), the higher rates of recharge can mobilise saline groundwater to the river and its floodplains, potentially resulting in deleterious impacts to downstream users and the environment. An extensive policy framework and management system is in place to manage the salinity risks [24]. The policy framework relies on an accounting system that assigns salinity effects (both debits and credits) to specific on-ground actions. Groundwater models, representing different irrigation districts, are used as the principal assessment tools to quantify groundwater returns and the salinity credits and debits resulting from irrigation practices.

2.2. Study Area Location, Climate, and Hydrology

The study area was the Loxton and Bookpurnong irrigation districts of South Australia (Figure 1). Annual rainfall is 275 mm and average maximum temperatures range from 15.6 °C in July to 31.3 °C in February. Irrigation occurs on the "highland" areas that flank the Murray River and its floodplain.

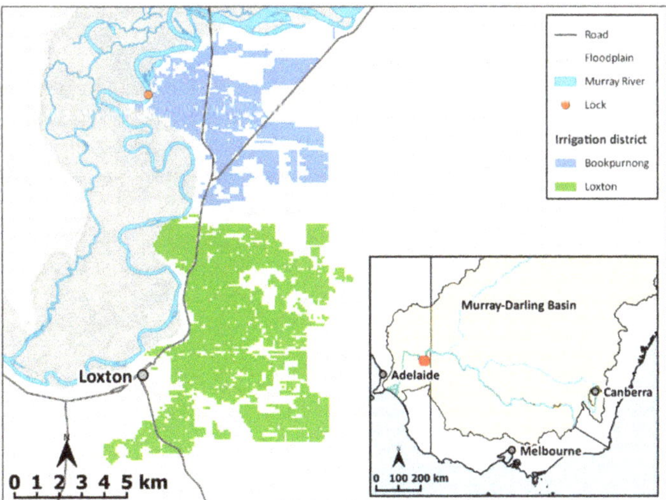

Figure 1. Location of the Loxton and Bookpurnong irrigation districts (the red marking on the inset map shows the location of the study area map).

The hydrology of the study area is dominated by the Murray River, which consists of a main river channel and a meandering network of anabranches and wetlands in its floodplain. There are no local tributaries and rainfall runoff is negligible. Flow in the Murray River is highly regulated by dams and storages upstream and by a series of weirs in its lower reaches that were constructed to aid navigation.

The weirs have created a regime of near-constant water levels, interspersed by floods that are irregular but large.

2.3. Landscape and Geology

The highland is composed of a dune-swale system, oriented east–west, with relatively subdued relief. It has been incised by the Murray River and its floodplain, which is about 20–30 m below the areas of irrigation.

The landscape is underlain by the Murray geological basin—a 550 m thick sequence of Cenozoic marine sediments. The Murray River has formed a trench through the upper sections of this sedimentary sequence. The floodplain trench is infilled by alluvial sediments to be level with marine Loxton Sands unit of the regional sedimentary sequence. The Loxton Sands are capped by the Blanchetown Clay, a widespread lacustrine deposit that underlies most, but not all, of the irrigation district. It is of variable depth and thickness, tending to occur from depths of 2–3 m below surface as a 5 m thick layer of laminated clays. It is overlain by the aeolian Woorinen Formation, which forms the dune-swale system at the surface and hosts sandy loam soils.

2.4. Hydrogeology

The regional water table occurs in Loxton Sands unit, which is in direct hydraulic connection with the alluvial floodplain aquifers that interact with the river. Groundwater in the Loxton Sands is highly saline (7000–50,000 mg/L). The Murray River is generally gaining along this reach, with groundwater (and salt) entering the river and floodplain by lateral flow from the Loxton Sands and some upwards leakage from deeper units in the Murray Basin [21].

In the areas of irrigation, the vadose zone is characterised by three layers in order of depth: loamy sands of the Woorinen Formation, the Blanchetown Clay (which may or may not be present), and the unsaturated upper sections of the Loxton Sands.

2.5. Irrigation and Drainage

The region is well suited to irrigated horticulture, having abundant sunshine, suitable soils, and access to surface water. Irrigation along the Murray River in South Australia commenced in the 1890s, with more intense development after the Second World War [25]. Most of the development in Loxton occurred in 1948 with the establishment of the Loxton Irrigation Trust, a government-sponsored initiative [26]. The Bookpurnong area was established more recently, from the 1980s onwards, as a private initiative.

The efficiency of irrigation has improved over time [25]. Prior to the 1970s, irrigation practices were either by flood or furrow irrigation with water supplied on a set roster. This has shifted progressively to sprinklers and drippers, and the pressurisation of off-take systems from the river. This has resulted in lower rootzone drainage rates relative to the application rates. The author of [26] estimated irrigation efficiency factors at decadal intervals since irrigation development. The estimates were based on a combination of anecdotal evidence from early time periods, as well as some selected irrigator surveys and benchmarking studies for particular crop types and application methods from later time periods [26]. The estimates are more accurate since 2000 but are uncertain for all time periods.

Rootzone waterlogging became evident soon after the development of irrigation, due the formation of perched water tables on the Blanchetown Clay. Sub-surface drainage systems were installed to manage the issue.

At Loxton, a comprehensive drainage scheme (CDS) was installed. This features a series of gravity flow pipelines that receive water from tile drainage networks on each property and delivers this water to a series of open bottom caissons, where water collects and is pumped to the floodplain for disposal. Data have been collected on the volumes of water intercepted. Drainage volumes increased until the late 1990s and have steadily declined since.

In Bookpurnong, drainage issues do not occur as commonly as at Loxton. About 30% of the irrigated area requires drainage and this is confined mostly to the southwest of the district. The drainage of these properties has been managed privately, in which a tile drainage network transfers water to sumps that are pumped from for disposal. There are no data available on the cumulative drainage volumes.

2.6. Groundwater Returns and Salinisation

Irrigation has caused recharge rates to increase substantially from background levels. Prior to European settlement, the region was covered by native Mallee vegetation that allowed very little water past its rootzone [27]. The clearing of native vegetation for dryland agriculture resulted in recharge rates increasing from 0.1 to 10 mm/year, and irrigation development has resulted in recharge rates exceeding 100 mm/year [28].

The higher recharge has led to the development of a groundwater mound within the Loxton Sands aquifer, resulting in increased salt loads to the river and the floodplain (see Figure 2). The lateral flow component from the Loxton Sands is thought to be the main mechanism for salt entering the floodplain and the river, and a network of dedicated pumping wells (a salt interception scheme, SIS) was installed in 2006 to actively reduce the groundwater returns to the river.

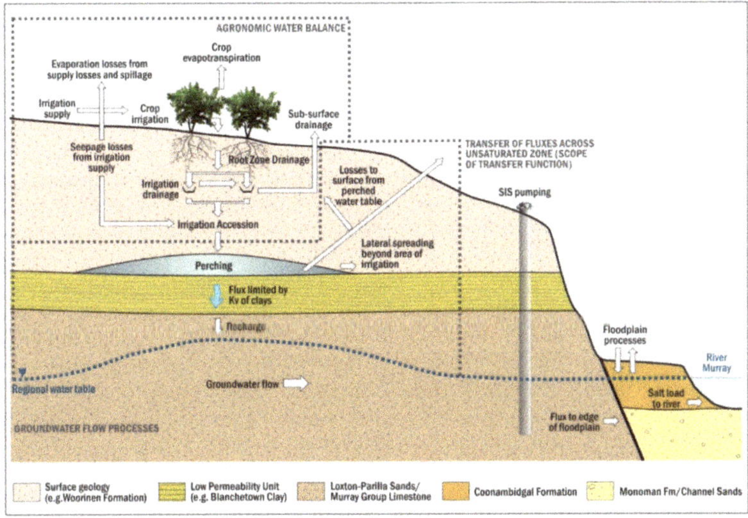

Figure 2. Conceptual model of groundwater returns to the Murray River from the Loxton and Bookpurnong irrigation districts, sourced with permission from [29]. Note that the Agronomic Water Balance used in this paper does not include irrigation drainage, which is addressed as an output of the transfer function.

As illustrated in Figure 2, the quantum of irrigation recharge is the key driving force for groundwater returns, and there are numerous processes that can influence it. These can be separated into three domains: (1) an agronomic water balance (AWB) that occurs at the surface/rootzone and defines the rootzone drainage rate; (2) the deeper vadose zone where variably saturated conditions, including perching, occur; and (3) the saturated zone.

2.7. Existing Methods to Model Groundwater Returns

The salinity management framework requires that groundwater returns (and their salt loads) to the river be quantified over history and into the future on the basis of past and present actions that include irrigation development, the implementation of efficiency measures, and SIS pumping. A 3D MODFLOW groundwater model [30] has been developed for this purpose and is updated over time to provide more refined salt load estimates and to account for any changes in management or observed data [21,31]. The domain of the groundwater model is primarily that of the saturated zone shown in Figure 2.

Estimates of rootzone drainage are obtained from an AWB undertaken on a district-wide scale at annual intervals [26]. Inputs to the water balance are based on measured rainfall and irrigation diversions. Outputs are based largely on an assumed irrigation efficiency factor, which represents the percentage to total water applied that is consumed by crop evapotranspiration. The residual of the water balance is the rootzone drainage rate. It is a comparatively small term relative to the other components (which have large error bands) and is therefore highly uncertain, particularly in early time periods for which there are limited data. It is also heavily dependent on the irrigation efficiency factor used in the calculation.

Although the estimates of rootzone drainage are available, they have not been used directly by the groundwater model. This is due largely to the lack of a simple, yet appropriate, vadose zone modelling tool at the time of the groundwater model development that could simulate the influence of perching of the transfer of rootzone drainage to recharge.

The recharge rates have been derived using the inverse method in which observed groundwater level fluctuations and aquifer pumping test data are used to constrain the hydrogeological parameters so that recharge may be derived inversely during model calibration, which was undertaken manually. The spatial zonation of recharge is based initially on the timing of irrigation development. The zonation evolves during the inversion process and is not highly constrained. The recharge rates over the whole irrigation areas are compared to the root zone drainage rates post-calibration as a check that the rates derived are comparable, but there is significant spatial variability between recharge zones. Other datasets are used, wherever possible, to constrain the model, such as under-river geophysics and river salinity data.

The calibrated groundwater model is then used to predict groundwater returns and salt loads into the future for a range of management scenarios. Under these scenarios, a subjective recharge rate of 100 mm/year has been selected to simulate ongoing irrigation.

The existing method is logical but has some significant limitations related to (1) the non-uniqueness of the recharge–aquifer parameter relationship and its associated uncertainty, and (2) the indirect link between on-ground actions and the derived recharge rates. These limitations result in a model that does not use all of the datasets available to constrain it, and one that relies on a subjective approach for the recharge rates used in predicting of groundwater returns (and salt loads). As such, the method is susceptible to bias.

3. Methods

3.1. Modelling Approach

The development of vadose zone transfer functions by [22,23] has allowed for the development of a more integrated modelling method. The method, illustrated in Figure 3, includes three elements that correspond to the three domains of the conceptual model (Figure 2):

1. The surface/rootzone is modelled by the AWB [26], which is applied spatially by using a systematic approach to recharge zonation, resulting in a time-varying rootzone drainage rate for each recharge zone.
2. The deeper vadose zone is modelled by the transfer functions (TF) developed by [22,23]. The rootzone drainage rates are passed through transfer functions the resulting in two outputs: a time-varying drainage rate representing water that is returned to the surface via perching, and a time-varying irrigation recharge rate.

3. The saturated zone is modelled by a groundwater flow model (a modified version of the existing Loxton–Bookpurnong model [21]) to generate groundwater returns and salt load to the river as an output.

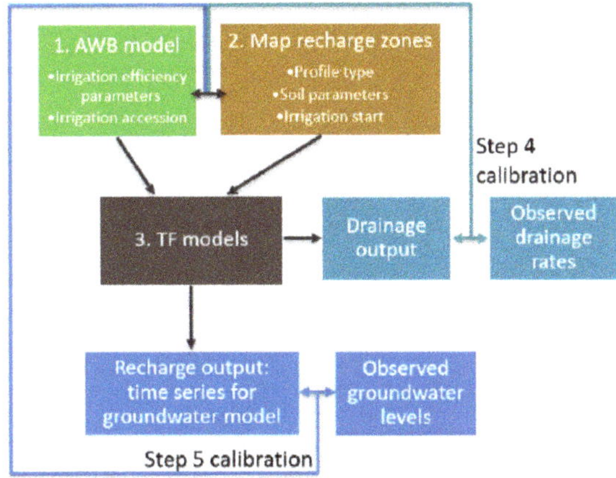

Figure 3. Process used in the integrated modelling method.

The method links on-ground actions directly with predicted groundwater returns and provides for two separate calibration targets: the observed data on drainage rates and groundwater levels.

3.2. Agronomic Water Balance

The AWB, as described by [26], was used to estimate the total rootzone drainage volume across each irrigation district. Unlike [26], who used the observed CDS drainage data as part of the water balance, it was removed from the calculations and set aside to use as a calibration target. The water balance was thus reduced to the following equation,

$$RZD = (P + I)(1 - Eff),\qquad(1)$$

where RZD is the root zone drainage volume across an entire irrigation district, P is the precipitation volume over the district (L^3/L^3), I is the total volume of irrigation water diverted to the district (L^3/L^3), and Eff is the irrigation efficiency factor (%), noting that other factors are included in the water balance during early time periods at Loxton to estimate seepage and transmission losses from open irrigation channels.

3.3. Recharge Zonation

Recharge zones were defined according the year of irrigation development (Figure 4a) and the nature of the vadose zone underlying each irrigation block (Figure 4b). The various combinations of these two factors resulted in 335 recharge zones across both irrigation districts (Figure 4c).

The year of irrigation development was obtained from historical aerial imagery of the irrigation footprint. The vadose zone was classified into units and subunits using the available borehole logs and geological models. The classification relates to the type of vadose zone transfer function required and its parameterisation. If the Blanchetown Clay was absent, a single layer (loam) transfer function was used and sub-units defined the varying thicknesses of the vadose zone. If the Blanchetown Clay was present, then a three-layer (loam–clay–sand) transfer function model was required, and sub-units define the varying layer thicknesses at different points in the landscape. The three-layer zones were also coded by whether a drainage network had been installed or not. This distinction can be used as a calibration target.

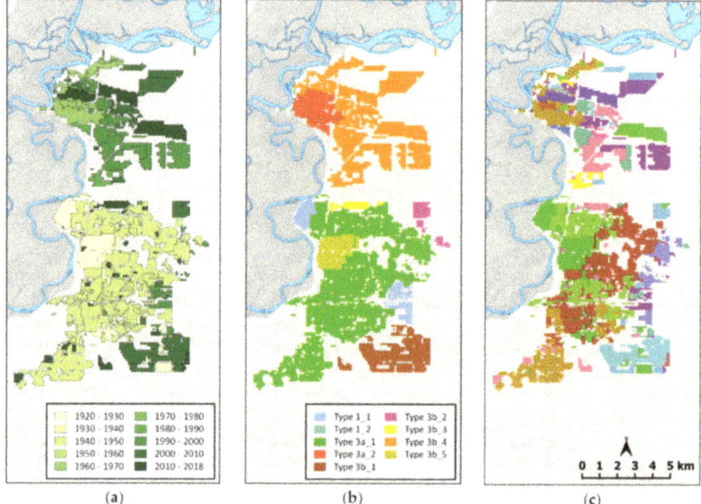

(a)   (b)   (c)

Figure 4. Method used to allocate recharge zones, showing (**a**) the timing of irrigation development, (**b**) classification of the vadose zone, and (**c**) resulting recharge zones

3.4. Transfer Functions

The TF models representing a one-layer or three-layer model are described by [22,23]. The one-layer model represents a vadose zone where the clay layer is absent and therefore perching and a requirement for drainage does not occur. The input is the time-varying RZD and the output is the time-varying irrigation recharge rate. The three-layer model was defined to represent a vadose zone where perching may occur due to the presence of a clay layer. The input is the time-varying RZD and the output is the time-varying irrigation recharge rate along with a time varying drainage rate. The drainage occurs when a perched water table reaches the surface of the model.

Both models are parameterized using the modified Maulem–van Genuchten function [32]. The parameters used were based on measured soil water retention data of Mallee soils [33].

A scripting process was developed to run the AWB, distribute RZD across the various recharge zones, and to call and run the TF models for each zone.

3.5. Partial Calibration of Irrigation Recharge

A partial calibration of the AWB-TF models can be performed using drainage data. This calibration can occur externally to any groundwater modelling, and several aspects of the drainage data can be used to constrain the model outputs, as follows:

- Where drainage infrastructure is absent, the modelled drainage should be negligible.
- Where drainage infrastructure is present, the modelled drainage should be non-negligible.
- Where there is observed drainage data, this modelled drainage should approximate these trends.

In calibrating the models at Loxton and Bookpurnong, it was considered that two main parameters could be varied as part of the calibration:

- The irrigation efficiency parameter (Eff) at any point in time, which controls the total recharge and drainage output provided by the model.
- The vertical hydraulic conductivity (K_v) of the Blanchetown Clay, which controls the partitioning between recharge and drainage—a lower K_v leads to more perching and more drainage, a higher K_v leads to less perching and more recharge.

3.6. Groundwater Model and Integration of Transfer Function

Various options can be employed to integrate the AWB and TF models with a groundwater model. In this study, a staged integration was undertaken to assess the merits of the approach. We considered a base case and two potential methods that integrate the AWB and TF models with the groundwater model focusing on different calibration approaches (Table 1).

Table 1. Methods tested for model integration.

Method	Recharge	Hydrogeological Parameters
LB2011	Inversely calibrated using groundwater data	Existing MODFLOW model
TF-A/B	AWB-TF, calibrated using drainage data	Recalibrated MODFLOW model
TF-C	AWB-TF-MODFLOW calibrated together using drainage and groundwater data	

The base case (LB2011) is the existing Loxton–Bookpurnong groundwater model in MODFLOW from which recharge rates are derived inversely via manual calibration.

TF-A/B uses drainage data to calibrate the AWB and TF models and derived recharge rates that are passed to the groundwater model. The MODFLOW model is recalibrated in accordance with the revised irrigation recharge and the observed groundwater level data. In this instance, the AWB-TF model was calibrated manually and the MODFLOW model was calibrated using an automated calibration procedure incorporating the Parameter ESTimation code, PEST [34].

TF-C calibrates the integrated (AWB-TF-MODFLOW) model simultaneously using drainage and groundwater data. In this instance, the integrated model was calibrated using PEST, but a manual calibration process could also be applied.

4. Results

4.1. Drainage and Irrigation Recharge Using Partial Calibration

An initial calibration of the AWB-TF models was undertaken using drainage data. It was found that the K_v of the clay layer was well constrained by the drainage targets and only minor adjustments were required from the default value of 0.03 cm/day (109.5 mm/year) (Table 2). Non-negligible drainage rates would occur in areas where drainage infrastructure is absent when the K_v was decreased substantially from the default value. Any substantial increases to the K_v were constrained by the observed drainage data at Loxton.

Table 2. Calibrated hydraulic conductivity (K_v) of Blanchetown Clay from use of drainage data.

Irrigation Area	Drainage Infrastructure	Blanchetown Clay K_v	
		Default Value	Calibrated Value
Bookpurnong	Present	0.03 cm/day (109.5 mm/year)	0.02 cm/day (73 mm/year)
	Absent		0.027 cm/day (98.6 mm/year)
Loxton	Present		0.02 cm/day (73 mm/day)
	Absent		0.03 cm/day (109.5 mm/day)

Figure 5 shows a comparison of the recharge and drainage rates derived from the existing methods and datasets to those derived by the new AWB-TF models developed. The AWB-TF models show recharge and drainage outputs that increased substantially in the late 1950s, reflecting the main period of irrigation development and accounting for a short time lag (<5 year), a period of stabilisation between 1960 and 2000, and a decline post-2000 as irrigation efficiency measures were introduced. The calibration focused on the period of late 1990s onwards, as this was where there was more confidence in the available datasets.

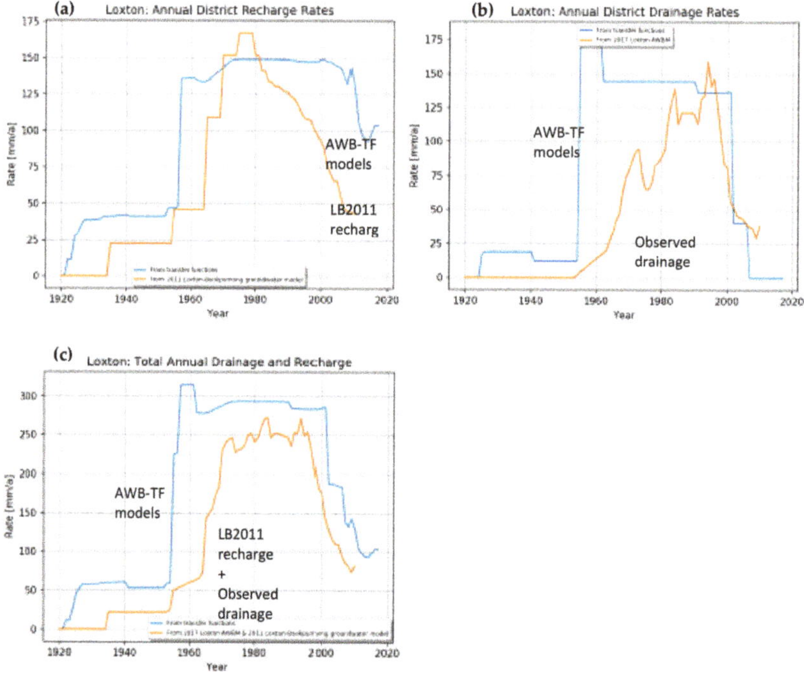

Figure 5. Comparison of recharge and drainage rates derived from existing methods and datasets (plotted in orange) to those derived the new agronomic water balance (AWB)-transfer functions (TF) models (plotted in blue) for the Loxton irrigation area, showing (**a**) techarge rates, (**b**) drainage rates, and (**c**) total drainage and recharge rates.

When recharge and drainage were combined, the outputs of the AWB-TF models compared reasonably to the combined LB2011 recharge and observed drainage rates (Figure 5c). However, there was divergence in the 1960–2000 period when recharge and drainage were considered separately. The LB2011 model had recharge declining from 1980 onwards at a period when drainage rates were increasing. This is counter-intuitive because if drainage rates were increasing then it would be expected that recharge was also increasing or was at least stable. It may be the result of the regional groundwater mound being intercepted by the drainage infrastructure and the LB2011 model needing to reduce recharge to account for this. Although this is possible, the LB2011 model does not model these processes explicitly, nor does its documentation describe these circumstances as the justification for reducing recharge over a period when drainage was increasing. In contrast, the AWB-TF model estimates gross recharge, and if the interception of the regional groundwater mound is occurring then it can be modelled explicitly by the groundwater model. As shown in this example, the new approach provides a much clearer representation of the irrigation actions affecting groundwater recharge, ultimately creating transparency in the estimation of salinity impacts to the river.

4.2. Comparison of Integrated Models

The integrated models obtained comparable calibration statistics to the existing method. The scaled residual mass statistic (SRMS) was 1.6% for the LB2011 model, 1.8% for TF-A/B, and 1.7% for TF-C. The calibrated horizontal hydraulic conductivities, K_h, of the regional water table aquifer are shown in Figure 6. The spatial distribution of K_h under TF-B was broadly similar to LB2011. Given that the conductivity field of TF-B was derived independently of LB2011, this provides some confidence in the new method.

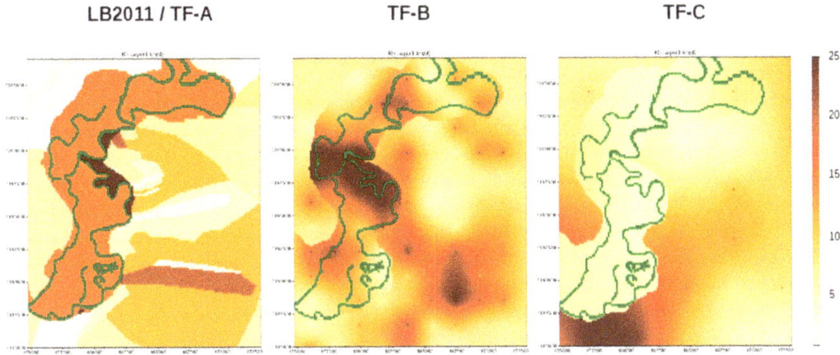

Figure 6. Comparison of calibration for horizontal hydraulic conductivity (m/d) of the Loxton Sands (regional water table aquifer) for the three models evaluated: LB2011/TF-A model, TF-B model, TF-C model.

In TF-C, a zone of high K_h was derived in the south of the model domain, which was not replicated by the other models. Closer inspection revealed a complication in this area of the model, with simultaneous calibration of drain conductance and efficiencies in the AWB, whereby greater recharge could be produced by the transfer function only to be offset by increases to drainage conductance in the MODFLOW model. This was a product of the automated calibration procedure, and further work is required to mitigate this effect, but this complication is not reflective of the integrated modelling approach being tested.

There were some key differences in how irrigation recharge was distributed in space and time. The inclusion of the AWB-TF components to the modelling tended to minimize the spatial variability in recharge rates, which could be considerable in the existing method (Figure 7). The recharge rates of the integrated models were also more stable temporally (see Figure 5a).

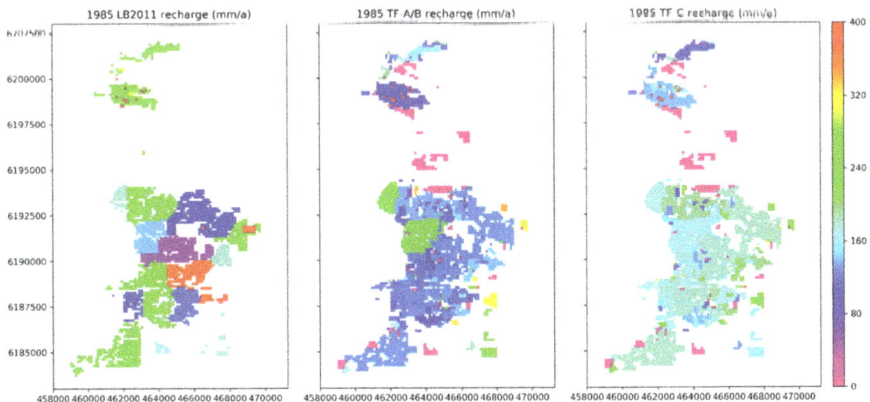

Figure 7. Distribution of irrigation recharge in 1985 within the existing LB2011 model, the TF-A/B model, and TF-C model.

Figure 8 compares the irrigation efficiency factor (*Eff*) used in the integrated models. In the TF-A/B model, *Eff* was taken from the default values of [26] and was not varied during the calibration, whereas in TF-C it was varied as part of the calibration. The recharge rates and estimated groundwater returns can be quite sensitive to *Eff*, and the large increase in the early 2000s was coincidental with the pressurisation of the irrigation supply system at Loxton.

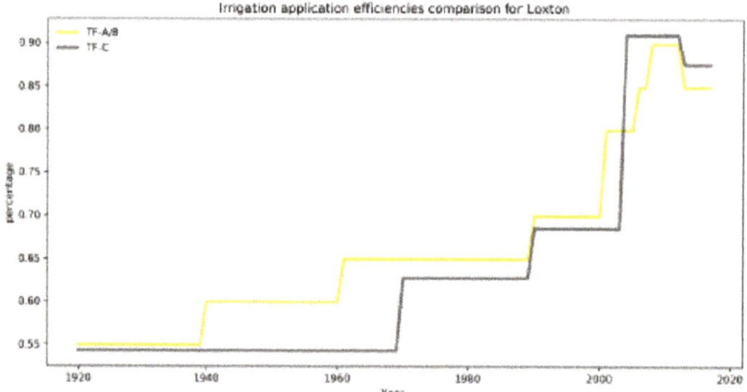

Figure 8. Irrigation efficiency factor used in the TF-A/B and TF-C models.

Figure 9 shows the projected salt flux to the river for the different models used. It is highest under the existing method.

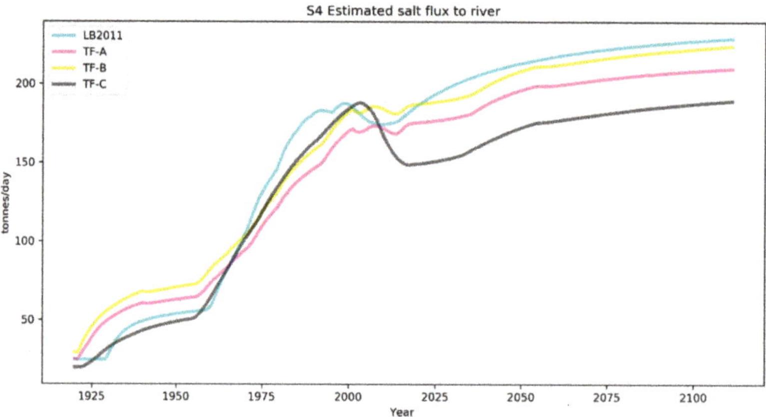

Figure 9. Estimated salt flux associated with groundwater returns to the river for the different models used. The modelled salt flux was determined using the calibrated models up until 2011. For LB2011, future projections were determined by assigning a constant irrigation recharge of 100 mm/year. For the other models, future projections were determined in accordance with the AWB and calibrated irrigation efficiency for 2011 and running this for 100 years through the TF models to generate a recharge rate into the future.

The integrated approach offers a transparent method of determining an appropriate recharge rate for estimating future salinity impacts. The basis of the method is to set the irrigation accession rate in accordance with the AWB and calibrated irrigation efficiency for 2011, and to run this into the future. By comparison, the existing methods subjectively select 100 mm/year as the recharge rate into the future on the basis of expert opinion rather than local datasets.

For TF-A/B, the continued *Eff* was 85% for both districts. For TF-C, the *Eff* was only adjusted slightly (87% for Loxton and 83% for Bookpurnong) but its sensitivity was such that small changes can result in large changes to the *RZD*. For instance, at Loxton, an adjustment in *Eff* from 85% to 83% resulted in a 13% decrease in the *RZD* (from 129 to 112 mm/year). This influenced the projected groundwater returns and shows the importance of having an objective procedure to calibrate *Eff*.

5. Discussion

5.1. Benefits of The Integrated Modelling Approach

For the irrigation areas with deep vadose zones in which perching can occur, the only viable approach before the development of the TF models was an inverse approach in which the aquifer conductivity values and then recharge values were inferred (with some limited adjustments of conductivity) on the basis of observed groundwater responses [21]. The integrated modelling approach clearly addresses a deficiency of the inverse approach, namely, not being able to transparently link actions to salt load impacts. In addition, the inverse method requires a complex process for calibrating recharge over many zones and stress periods, and may have few other datasets to constrain what is a non-unique problem. A careful logic and many groundwater model runs are required to provide sensible values. The integrated modelling approach is simpler than this because it is objectively formulated.

The results demonstrate the impact that recharge input values in a groundwater model has on the salt flux to river estimates used for register entries. Given a scenario where there is limited data to constrain recharge values, use of the integrated method provides some means to quantify the reduction in uncertainty of recharge estimates via automated calibration software. There is presently no method to fully quantify reductions to uncertainty using manual calibration. Adopting the transfer function for future assessment promotes transparency and repeatability of the modelling approach. Moreover, it will promote consistency between groundwater models used for other irrigation areas by reducing subjectivity of the calibration process.

5.2. Limitations of the Integrated Modelling Approach

The trial demonstrated that the integrated model may require an increase in model development time. The extent of the increase will be dependent on the modeller's familiarity with a scripting language such as Python and/or an automated calibration utility such as PEST.

Initially, the method will require time and effort to upskill users and to incorporate routinely into current workflow. However, this additional cost will reduce over time once the approach and the linking script are developed for a particular model.

A lack of data may still affect the success of applying the integrated approach. Drainage data for the Loxton district provided some constraint on the outputs of both the AWB and transfer function in this trial. In addition, attempts to minimize the difference between simulated drainage in the AWB and transfer function provided a further constraint on the calibration of irrigation application efficiencies and clay conductance.

Access to high performance computing is also strongly recommended if the integrated approach is to be used with automated calibration.

5.3. Recommendations

Despite the success of the trial implementing the integrated modelling approach, further areas of investigation are required and include:

- Further testing of ways to optimise the AWB and transfer function parameters to produce recharge trends that better match observed head variation patterns. Use of more discrete time periods within the AWB is one option.
- The development of scripting processes to simulate the deactivation of irrigation blocks.
- Development of transfer function models for situations where there are multiple clay layers in the vadose zone.
- Testing the integrated method in situations where perching occurs but the vadose zone is shallow. In these settings, a forward modelling method is typically used [19,20], but there is no feedback loop and use of drainage data to constrain the AWB parameters.

6. Conclusions

Irrigation areas have a complex history (spatially and temporally) with respect to the timing of irrigation development, water use efficiency, drainage, climate, and crop type, among other factors. This complexity can lead to a lack of transparency, obscure interpretation, and require additional resources and skills to represent it appropriated within a modelling framework. Any new methodology has to ensure that it does not add to the complexity of prior approaches and it must be widely applicable. Our development of the transfer function and integrated modelling was cognisant of these considerations. It attempted to simulate the transmission of the unsaturated zone with as few parameters as possible. It implemented a flexible method that can be applied in other semi-arid environments where there is a deep vadose zone. It makes use of existing datasets and attempts to use all of these datasets more thoroughly to calibrate and constrain the model outputs.

This paper described an integrated methodology for the estimation of groundwater returns to streams from irrigation areas with deep vadose zones and perched water tables. This methodology was implemented for an irrigation district in south-eastern Australia, adjacent to the lower reaches of the River Murray. An existing surface water balance model and the spatial distribution of the historical irrigation development were used to assess irrigation accessions to the unsaturated zone. A recently developed unsaturated zone model [22,23] estimated the groundwater recharge from the irrigation accession. Two different implementations of a groundwater model using these data were compared to a pre-existing groundwater model by using recharge calibrated using groundwater responses. In the first implementation, the newly estimated recharge values replaced the recharge values in the pre-existing model, that is, the saturated zone properties were the same. In the second implementation, a new set of saturated zone parameters were calibrated.

The study has shown that

- The implementation of the integrated methodology requires some additional data (drainage information, high quality maps of irrigation development, soil maps). However, the pilot study showed that many of these could be either accessed or generated within a constrained time.
- The drainage information could be used to constrain the main soil parameters.
- The irrigation accession, unsaturated zone model, and the recharge data layers were linked through Python scripting. Due to variations between irrigation districts and available data, some changes in the script will be generally be required.
- Recharge data generated by the unsaturated zone model can readily replace those generated by the arduous inversion process, while providing greater transparency.
- The calibration of aquifer conductivity, using the new recharge data, generates credible parameters, even without the use of pump tests and other independent information.

Further work is required to optimize the calibration methodology, standardise scripting approaches, and test the method in other vadose zone settings.

Author Contributions: Conceptualization, G.W.; data curation, D.C., T.L., and K.B.; formal analysis, D.C., T.L., G.W., and T.S.; funding acquisition, D.C. and J.W.; investigation, D.C., T.L., G.W., and T.S.; methodology, D.C., T.L., and G.W.; project administration, D.C.; software, T.L., G.W., and T.S.; supervision, G.W. and J.W.; writing—original draft, D.C.; writing—review and editing, G.W. and J.W. All authors have read and agreed to the published version of the manuscript.

Funding: The authors would like to thank the Murray-Darling Basin Authority (MDBA) and the Department of Environment and Water, South Australia (DEW), for funding support.

Acknowledgments: The authors would like to thank and acknowledge Emmanuel Xevi for his support throughout this work program, and Hugh Middlemis, Ray Evans, and Prathapur for technical advice.

Conflicts of Interest: The authors declare no conflict of interest.

References

1. Wang, Q.J.; Walker, G.; Horne, A. *Potential Impacts of Groundwater Sustainable Diversion Limits and Irrigation Efficiency Projects on River Flow Volume under the Murray-Darling Basin Plan.: An. Independent Review*; Report Written for the MDBA; University of Melbourne, MDBA: Canberra, Australia, 2018; 73p.
2. Perry, C. Efficient irrigation; inefficient communication; flawed recommendations. *Irrig. Drain.* **2007**, *56*, 367–378. [CrossRef]
3. Meals, D.W.; Dressing, S.A.; Davenport, T.E. Lag time in water quality response to best management practices: A review. *J. Env. Q.* **2010**, *39*, 85–96. [CrossRef] [PubMed]
4. Williams, J.; Grafton, R.Q. Missing in action: Possible effects of water recovery on stream and river flows in the Murray–Darling Basin, Australia. *Australas. J. Water Resour.* **2019**, *23*, 78–87. [CrossRef]
5. Cook, P.G.; Jolly, I.D.; Walker, G.R.; Robinson, N.I. From drainage to recharge to discharge: Some timelags in subsurface hydrology. *Developments in Water Science* **2003**, *50*, 319–326.
6. Khan, S.; Rana, T.; Carroll, J.; Wang, B. *Managing Climate, Irrigation and Ground Water Interactions using a Numerical Model.: A Case Study of the Murrumbidgee Irrigation Area*; CSIRO Land and Water Technical Report No. 13/04; CSIRO: Canberra, Australia, 2004.
7. Kurtzman, D.; Scanlon, B.R. Groundwater Recharge through Vertisols: Irrigated Cropland vs. Natural Land, Israel. *Vadose Zone J.* **2011**, *10*, 662–674. [CrossRef]
8. Besbes, M.; de Marsily, G. From infiltration to recharge: Use of a parametric transfer function. *J. Hydrol.* **1984**, *74*, 271–293. [CrossRef]
9. Mattern, S.; Vanclooster, M. Estimating travel time of recharge water through a deep vadose zone using a transfer function model. *Environ. Fluid Mech.* **2010**, *10*, 121–135. [CrossRef]
10. Wang, X.S.; Ma, M.-G.; Li, X.; Zhao, J.; Dong, P.; Zhou, J. Groundwater response to leakage of surface water through a thick vadose zone in the middle reaches area of Heihe River Basin, in China. *Hydrol. Earth Syst. Sci.* **2010**, *14*, 639–650. [CrossRef]
11. Bouwer, H. Effect of Irrigated Agriculture on Groundwater. *J. Irr. Drain. Eng.* **1987**, *113*. [CrossRef]
12. Orr, B.R. *A transient numerical simulation of perched groundwater flow at the test reactor area, Idaho National Engineering and Environmental Laboratory, Idaho, 1952–1994*; U.S. Geological Survey: Reston, VA, USA, 1999; pp. 99–4277.
13. Truex, M.J.; Oostrom, M.; Carroll, K.C.; Chronister, G.B. *Perched-Water Evaluation for the Deep Vadose Zone Beneath the B, BX, and BY Tank Farms Area of the Hanford Site*; Pacific Northwest National Laboratory: Richland, WA, USA, 2013; p. 99352.
14. Oostrom, M.; Truex, M.J.; Carroll, K.C.; Chronister, G.B. Perched-water analysis related to deep vadose zone contaminant transport and impact to groundwater. *J. Hydrol.* **2013**, *505*, 228–239. [CrossRef]
15. Weiss, M.; Gvirtzman, H. Estimating Ground Water Recharge using Flow Models of Perched Karstic Aquifers. *Groundwater* **2007**, *45*, 761–773. [CrossRef] [PubMed]
16. Jolly, I.D.; Cook, P.G.; Allison, G.B.; Hughes, M.W. Simultaneous water and solute movement through unsaturated soil following an increase in recharge. *J. Hydrol.* **1989**, *111*, 391–396. [CrossRef]
17. Rossman, N.R.; Zlotnik, V.A.; Rowe, C.M.; Szilagyi, J. Vadose zone lag time and potential 21st century climate change effects on spatially distributed groundwater recharge in the semi-arid Nebraska Sand Hills. *J. Hydrol.* **2014**, *519*, 656–669. [CrossRef]
18. Wang, L.; Wu, J.Q.; Hull, L.C.; Schafer, A.L. Modeling Reactive Transport of Strontium-90 in a Heterogeneous, Variably Saturated Subsurface. *Vadose Zone J.* **2010**, *9*, 670–685. [CrossRef]
19. Kendy, E.; Bredehoeft, J.D. Transient effects of groundwater pumping and surface-water-irrigation returns on streamflow. *Water Resour. Res.* **2006**, *42*. [CrossRef]
20. Aquaterra. *Sunraysia Sub-Regional Groundwater Flow Model (EM2.3). Final Report*; Report Prepared for Goulburn-Murray Water and Murray-Darling Basin Authority; MDBA: Canberra, Australia, 2009.
21. Yan, W.; Li, C.; Woods, J. *Loxton–Bookpurnong Numerical Groundwater Model*; DFW Technical Report 2011/22; Government of South Australia: Adelaide, Australia, 2011; Volume 1: Report and Figures.
22. Walker, G.R.; Currie, D.; Smith, T. Modelling recharge from irrigation developments with a perched water table and deep unsaturated zone. *Water* **2020**. (under review).
23. Walker, G.R.; Currie, D.; Smith, T. Modelling the effect of efficiency measures and increased irrigation development on groundwater recharge through a deep vadose zone. *Water* **2020**. (under review).

24. Murray-Darling Basin Authority. *Basin Salinity Management 2030, BSM2030*; MDBA Publication: Canberra, Australia, 2015.

25. Adams, A.; Meissner, T. *How Efficient Are We? A report on Water Use Efficiency in the South Australian Murray-Darling Basin*; Department of Water, Land and Biodiversity Conservation, Government of South Australia: Adelaide, Australia, 2010.

26. Meissner, T. *Estimation of Accession for Loxton and Bookpurnong Irrigation Areas*; Report prepared for the Department for Environment and Water; Government of South Australia: Adelaide, Australia, 2019.

27. Allison, G.B.; Cook, P.G.; Barnett, S.R.; Walker, G.R.; Jolly, I.D.; Hughes, M.W. Land clearance and river salinization in the western Murray Basin, Australia. *J. Hydrol.* **1989**, *119*, 1–20. [CrossRef]

28. Cook, P.G.; Leaney, F.W.; Jolly, I.D. *Groundwater recharge in the Mallee Region, and salinity implications for the Murray River*; CSIRO Land and Water Technical Report, 45/01; CSIRO: Clayton, VIC, Australia, 2001.

29. Walker, G.R.; Currie, D.; Smith, T.; Middlemis, H. *Development of a transfer function to model irrigation recharge*; Report prepared for the Murray-Darling Basin Authority; MDBA: Canberra, Australia, 2018.

30. McDonald, M.G.; Harbaugh, A.W. A modular three-dimensional finite-difference groundwater flow model. Techniques of Water-Resources Investigations of the United State Geological Survey. In *Modelling Techniques—Book 6*; United States Geological Survey; McDonald, M.G., Harbaugh, A.W., Eds.; Reston, VA, USA, 1988.

31. Yan, W.; Howles, S.; Hill, A. *Loxton Numerical Groundwater Model. 2004*; DWLBC Report Book 2005/16; Department of Water, Land and Biodiversity Conservation, Government of South Australia: Adelaide, Australia, 2005.

32. Schaap, M.G.; van Genuchten, M.T. A Modified Mualem–van Genuchten Formulation for Improved Description of the Hydraulic Conductivity Near Saturation. *Vadose Zone J.* **2006**, *5*, 27–34. [CrossRef]

33. Meissner, T. Relationship between soil properties of Mallee soils and parameters of two moisture characteristics models. In Proceedings of the 3rd Australian and New Zealand Soils Conference (SuperSoil 2004), Sydney, Australia, 5–9 December 2004.

34. Doherty, J. *Calibration and Uncertainty Analysis for Complex Environmental Models*; Watermark Numerical Computing: Brisbane, Australia, 2015.

Article

Using Wastewater in Irrigation: The Effects on Infiltration Process in a Clayey Soil

Ammar A. Albalasmeh [1],*, Mamoun A. Gharaibeh [1], Ma'in Z. Alghzawi [1], Renato Morbidelli [2], Carla Saltalippi [2], Teamrat A. Ghezzehei [3] and Alessia Flammini [2]

[1] Department of Natural Resources and The Environment, Faculty of Agriculture, Jordan University of Science and Technology, Irbid 22110, Jordan; mamoun@just.edu.jo (M.A.G.); maen_alghazzawi.just@live.com (M.Z.A.)

[2] Department of Civil and Environmental Engineering, University of Perugia, 06125 Perugia, Italy; renato.morbidelli@unipg.it (R.M.); carla.saltalippi@unipg.it (C.S.); alessia.flammini@unipg.it (A.F.)

[3] Life and Environmental Sciences, University of California, Merced, CA 95343, USA; taghezzehei@ucmerced.edu

* Correspondence: aalbalasmeh@just.edu.jo; Tel.: +962-2-7201000 (ext. 22050)

Received: 13 January 2020; Accepted: 27 March 2020; Published: 29 March 2020

Abstract: Soil water infiltration is a critical process in the soil water cycle and agricultural practices, especially when wastewater is used for irrigation. Although research has been conducted to evaluate the changes in the physical and chemical characteristics of soils irrigated by treated wastewater, a quantitative analysis of the effects produced on the infiltration process is still lacking. The objective of this study is to address this issue. Field experiments previously conducted on three adjacent field plots characterized by the same clayey soil but subjected to three different irrigation treatments have been used. The three irrigation conditions were: non-irrigated (natural conditions) plot, irrigated plot with treated wastewater for two years, and irrigated plot with treated wastewater for five years. Infiltration measurements performed by the Hood infiltrometer have been used to estimate soil hydraulic properties useful to calibrate a simplified infiltration model widely used under ponding conditions, that were existing during the irrigation stage. Our simulations highlight the relevant effect of wastewater usage as an irrigation source in reducing cumulative infiltration and increasing overland flow as a result of modified hydraulic properties of soils characterized by a lower capacity of water drainage. These outcomes can provide important insights for the optimization of irrigation techniques in arid areas where the use of wastewater is often required due to the chronic shortage of freshwater.

Keywords: treated wastewater; irrigation techniques; infiltration modeling

1. Introduction

A quantitative study of water movement in the vadose zone allows us to identify strategies for water conservation, flood/runoff and erosion control, and the assessment of aquifer potential contamination due to migration of water-soluble chemicals [1]. Infiltration of irrigation water is one of the most critical processes for successful agricultural activities [2]. It is a key dynamic process to be considered for the design of irrigation systems and optimization, irrigation scheduling, and irrigation management [3]. This process assumes much more importance in arid and semi-arid regions where, because of short periods and low amounts of rainfall, water is a scarce resource considered as a limiting factor for agricultural production. In these areas, the chronic water shortage has compelled the decision-makers to look for non-conventional water sources for irrigation. One of these is the treated wastewater (TWW) [4], which also gives the advantage of low cost if compared with other solutions such as seawater. In this context, a continuous monitoring of the TWW and soil parameters is required to guarantee a sufficient level of water and soil quality for efficient plant development. In comparison to freshwater, usually, TWW has a higher content of organic matter and nutrients that, particularly in arid soils, are required for plant growth. However, it contains some elements that can adversely affect soil and plant [5]. Major effects produced by TWW are the physical clogging by suspended solids and the bioclogging facilitated by dissolved organic matter [6,7]. The clogging process typically results in the reduction of soil porosity and potential hydraulic conductivity. Bedbabis et al. [6] reported a decrease in soil infiltration after 4 years of using treated wastewater in irrigation. Similar results were obtained by Alizadeh et al. [8] in Iran as a result of using treated wastewater for irrigation of a cornfield for 2 years where the infiltration rate decreased by 15.6%. Moreover, Tunc and Sahin [7] reported a decrease in soil infiltration after having used treated wastewater in irrigating different crops grown in loamy soil, as a result of decreasing macropores by the suspended materials in the TWW.

On these bases, it is clear that a recommended use of TWW as the irrigation water source requires complete knowledge of its long-term effects on both hydraulic characteristics and the quality level of agricultural soils. Recently, Gharaibeh et al. [9] investigated the long-term impacts of irrigation with TWW on the physicochemical properties of soil through multi-year field trials. Infiltration measurements on three plots subjected to different irrigation durations were involved. They found that irrigation with TWW for a few years affected soil physicochemical properties producing an increase of electrical conductivity and sodium adsorption ratio. On the other hand, a decrease in pH, infiltration and unsaturated hydraulic conductivity due to pore-clogging by surface deposition of suspended materials was also observed. Furthermore, a slight decrease in bulk density and observable increases in aggregation percentage due to a significant increase in the organic matter were also highlighted. However, in this study, the effects of physical and chemical modifications of soil on the soil capability of absorbing water and generating surface runoff were not quantitatively extrapolated.

The objective of this work is to integrate the analysis by Gharaibeh et al. [9] through a quantitative estimate of the expected effects on the infiltration and runoff production processes due to multi-year TWW irrigation. This issue represents an open challenge in the light of optimizing irrigation techniques in arid and semi-arid geographic zones. Field experiments previously conducted on three adjacent field plots characterized by the same clayey soil but subjected to three different irrigation treatments have been used to address this issue. The three irrigation conditions were: non-irrigated (natural conditions) plot, irrigated plot with TWW for two years, and irrigated plot with TWW for five years. Data of infiltration measurements, earlier performed by the Hood infiltrometer, have been used here to estimate the soil hydraulic properties. These quantities have enabled us to calibrate a well-known infiltration model under the ponded conditions representative of the irrigation stage. The adopted approach has provided insights that can be useful in irrigation system designing as well as in optimizing the use of TWW.

2. Materials: Field Experiments

2.1. Experimental Site

The field experiments used here and previously described by Gharaibeh et al. [9] were conducted at Jordan University of Science & Technology (JUST) (32°27′57.4″ N latitude, 35°57′54.4″ E longitude), 70 km north of Amman, Jordan. The experiments were performed on three adjacent field plots (each of 0.8 ha) characterized by the same soil texture. The soil is classified as fine and mixed with clayey soil texture (clay 48%, silt 37%, and sand 15%) and thermic Typic Calcixerert characterized by 15% $CaCO_3$ content. The three plots were subjected to three different TWW irrigation patterns. The first plot was not-irrigated (rain-fed) and is used here as the no-TWW benchmark. The second plot was irrigated with TWW for two years, while the third plot was irrigated with TWW for five years. The TWW used in the experiments was supplied from a wastewater treatment plant located in the JUST campus, which uses rotating biological contactors. More details about water characteristics and irrigation strategies can be found in Gharaibeh et al. [9]. In the following sections, these treatments are referred to as 0 YR,

2 YR, and 5 YR, respectively. The main chemical soil characteristics of the three plots are presented in Table 1.

Table 1. Selected chemical properties for the soil of the three plots: pH, electrical conductivity (EC), organic matter (OM), and cation-exchange capacity (CEC).

Treatment/Plot	pH	EC (dS m^{-1})	OM (%)	CEC (cmole$_{(+)}$ kg^{-1})
0YR	6.9	0.7	2.77	32.49
2YR	7.7	1.68	4.37	31.16
5YR	7.4	2.09	7.19	33.44

2.2. Measurements of Hydraulic Properties

The Hood infiltrometer (IL-2700, Umwelt-Gerate-Technik GmbH, Muncheberg, Germany) was used for in-situ infiltration measurements following Schwarzel and Punzel [10]. This device was chosen because it maintains the soil surface and pore system undisturbed. Furthermore, it allows us to perform infiltration measurements at different water tensions. The measurements were realized in 2013 on an undisturbed soil surface. Each experiment was performed at least three days after the last rain or irrigation event. Five measurements (replicates) per each site (treatment) were conducted on randomly distributed locations. Infiltration tests were carried out by applying on the soil surface tensions ranging from 0 mm to the value of the soil bubbling point with steps of 20 mm. For each tension value, the infiltration rate was allowed to reach steady conditions for approximately 8 min before the tension level was changed to the next level.

Following Gharaibeh et al. [9], steady-state infiltration can be described using Wooding model [11]:

$$q_s(\Psi) = K(\Psi)\left(1 + \frac{4}{\pi r \alpha}\right) \tag{1}$$

where $q_s(\Psi)$ is the steady infiltration rate at the fixed tension Ψ, $K(\Psi)$ is the unsaturated hydraulic conductivity, r is the infiltrometer radius and α is the sorptive number. Substituting the exponential model of Gardner [12]:

$$K(\Psi) = K_s e^{\alpha \Psi} \tag{2}$$

and applying the natural logarithm to both sides, Equation (1) becomes:

$$\ln\left[q_s(\Psi)\right] = \alpha \Psi + \ln\left[K_s\left(1 + \frac{4}{\pi r \alpha}\right)\right] \tag{3}$$

where K_s is the saturated hydraulic conductivity. Equation (3) highlights a linear relationship between $\ln[q_s(\Psi)]$ and Ψ with α representing the slope. Linear regression can be therefore used on experimental pairs of $\ln[q_s(\Psi)]$ and Ψ for estimating α. The estimate of K_s can be then obtained by Equation (1) for $\Psi = 0$:

$$K_s = \frac{q_s(0)}{1 + (4/\pi r \alpha)} \tag{4}$$

Figure 1 shows an example of results based on the experiments performed on the three plots and described by Gharaibeh et al. [9]. Specifically, Figure 1a shows the instantaneous infiltration curves obtained during the first replicate of the infiltration experiment in each of the three plots subjected to different irrigation treatments (0YR, 2YR, and 5YR). Figure 1b represents cumulative infiltration curves averaged over the five replicates performed in each plot. As can be seen, the cumulative infiltration at the end of the 90 min experiments for the 0YR, 2YR, and 5YR treatments were 140 mm, 68 mm, and 59 mm, respectively, showing a significant reduction of soil infiltration capacity when TWW is applied. This reduction is likely due to the high load of the suspended solids present in the treated wastewater [9,13].

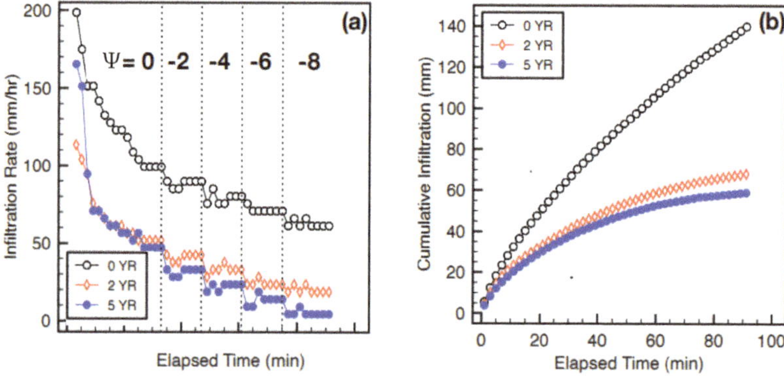

Figure 1. (**a**) Instantaneous infiltration rates associated with the first replicate; (**b**) mean cumulative infiltration curves obtained on the three plots with different earlier irrigation treatments (0YR, 2YR, and 5YR) [9].

Through the procedure described in Section 2.2, Gharaibeh et al. [9] obtained the saturated hydraulic conductivity and sorptive number averaged over five replicates of infiltration measurements realized on each plot/treatment (Table 2). Based on these results K_s is not significantly different among the three considered treatments even though a slightly decreasing trend can be detected with increasing the number of TWW irrigation years. This result suggests that the usage of TWW in irrigation does not alter morphology and connectivity of the largest pores which mainly influences the saturated hydraulic conductivity. However, the estimate of the sorptive number shows a relevant difference in the three plots with values significantly increased for TWW irrigated sites where, as a consequence, the unsaturated hydraulic conductivity expressed through the Gardner [12] model had lower values (see Figure 2). Considering that the sorptive number parameter indicates the relative magnitudes of gravity and capillarity forces during unsaturated flow [14], this outcome suggests a significant reduction of fine pores, that drain water at suction levels < 0 cm, with respect to the total porosity. This evidence was justified by Gharaibeh et al. [9] also with the presence in TWW of both high loads of organic material and suspended solids that tend to settle in the finer soil pore spaces where the flow velocity is lower. Furthermore, the application for long periods of wastewater determined a reduction and disconnection of soil micro- and mesopores leading to a significant drop in hydraulic conductivity of unsaturated soils.

Table 2. Saturated hydraulic conductivity, K_s, and sorptive number, α, estimated through the procedure described in Section 2.2 and averaged on five replicates of the infiltration tests performed on each plot [9].

Treatment/Plot	K_s (cm/h)	α (1/cm)
0YR	2.94	0.056
2YR	2.75	0.161
5YR	2.69	0.212

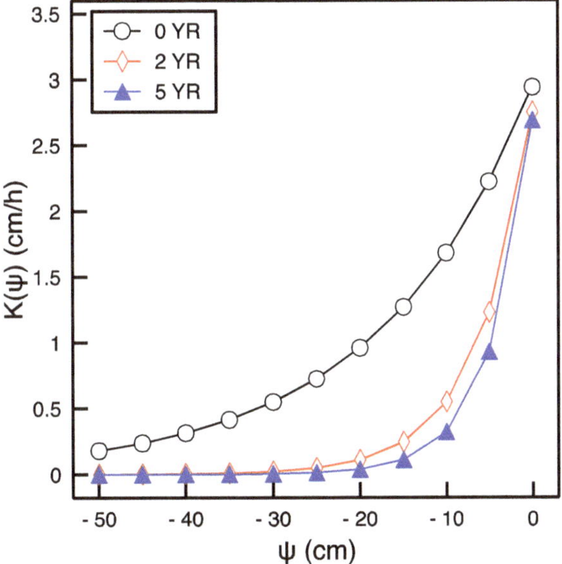

Figure 2. Unsaturated hydraulic conductivity, K, as a function of tension, Ψ, according to the Gardner (1958) model estimated for the three experimental plots with different irrigation treatments (0YR, 2YR, and 5YR).

3. Methodology

In this study, a simplified infiltration model widely used under ponding conditions was selected [15] This model is based on an infinite-series solution of the Richards flow equation [16] under the hypothesis of ponding conditions, which results in a two-term infiltration equation expressed as:

$$I = St^{1/2} + At \tag{5}$$

where I is the cumulative infiltration, t is the time, S is the soil sorptivity, and A is a constant that is related to the saturated hydraulic conductivity (typically assumed as 0.4 Ks).

The sorptivity parameter can be defined as

$$S(\Psi_0) = \lceil \gamma(\theta(\Psi_0) - \theta(\Psi_i)) \int_{\Psi_i}^{\Psi_0} K(\Psi)\,d\Psi \rceil^{\frac{1}{2}} \; \Psi_i < \Psi_0 < 0, \; \theta(\Psi_i) = \theta_i < \theta(\Psi_0) < \theta_s \tag{6}$$

where θ is the soil volumetric water content, the subscripts i and s refer to initial and saturated conditions, respectively, and γ =1.818 is a dimensionless empirical constant [17] related to the shape of the wetting front. Following Reynolds and Clarke Topp [18] and using Equation (2), the following equation holds:

$$S(\Psi_0) = \lceil \gamma(\theta(\Psi_0) - \theta(\Psi_i)) \frac{K(\Psi_0)}{\alpha(\Psi_0)} \rceil^{\frac{1}{2}} \tag{7}$$

This equation highlights that sorptivity reduces with increasing antecedent water content, decreasing hydraulic conductivity and increasing sorptive number. For $\Psi_0 = 0$ Equation (7) becomes:

$$S = \lceil \gamma(\theta_s - \theta_i) \frac{K_s}{\alpha} \rceil^{\frac{1}{2}} \tag{8}$$

Based on Equation (8) values of sorptivity have been estimated for the three plots/treatments of this study (Table 3).

Table 3. Philip's model parameters estimated for the three experimental plots of this study. The difference between saturated soil water content, θ_s, and initial soil water content, θ_i, is also given.

Treatment/Plot	$\theta_s - \theta_i$	S (cm/h$^{0.5}$)	A (cm/h)
0 YR	0.49	6.84	1.18
2 YR	0.42	3.61	1.10
5 YR	0.41	3.08	1.08

4. Results and Discussion

Based on the Philip model parameters estimated for the three experimental plots/treatments and shown in Table 3, it can be deduced that the sorptivity markedly decreases with increasing the period of irrigation using TWW. On the other hand, the A parameter is rather similar for all treatments. These results suggest that A, which is related to the saturated hydraulic conductivity and connectivity of the largest pores, is not affected by the TWW movement. In contrast, the decrease of S could be due to the clogging of the small pores being this parameter mainly influenced by the sorptive number for invariant values of K_s (Equation (8)). In this context small differences of antecedent soil conditions in terms of $\theta_s - \theta_i$—slightly decreasing from 0YR to 5YR treatments—were observed among the three plots (see Table 3).

Adopting the parameters of Table 3, the Philip model has been applied in the case of an irrigation process of duration 1.5 h and rate sufficiently high to determine quickly the saturation of soil surface in the three plots with different earlier irrigation treatments. The results obtained are shown in Figure 3 in terms of cumulative infiltration. The simulated cumulative infiltrations (Figure 3) at the irrigation end for the 0YR, 2YR, and 5YR plots were 101, 61, and 54 mm, respectively. Statistical analysis, involving one-way ANOVA test at the probability level $p < 0.001$ and Tukey's Honestly Significant Difference test at level $p < 0.05$, has shown that there are no significant differences between the two plots irrigated with TWW (2YR and 5YR treatments), while the cumulative infiltration in these two plots is significantly lower than that of the control plot (0YR). This significant decrease in cumulative infiltration in the 2YR and 5YR TWW irrigated soils could be explained by the high load of suspended solids present in the TWW. When soil is irrigated by TWW, these suspended materials settle in the smaller pores. With time, the micropores as well as the mesopores become smaller and disconnected producing a significant reduction in infiltration rate and cumulative infiltration. In agreement with this conclusion, Viviani and Iovino [19] showed a reduction in soil porosity which led to a decrease in soil infiltration rate as a result of pore-clogging due to the use of TWW. Similarly, Bardhan et al. [13] concluded that suspended solids loaded in the TWW reduced soil infiltrability due to pore-clogging.

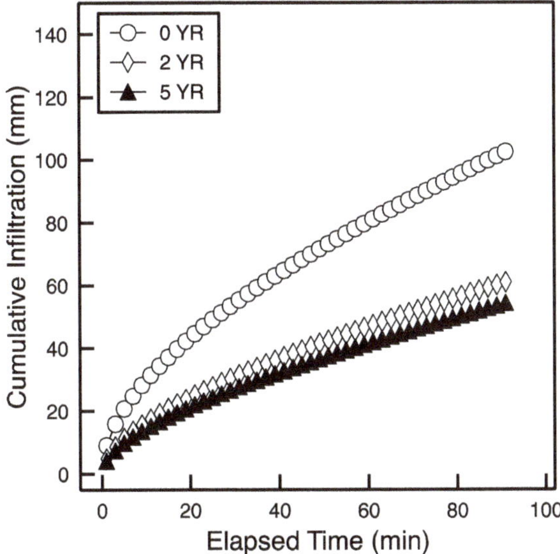

Figure 3. Cumulative infiltration curves obtained by Philip's model for the three experimental plots subjected to different irrigation treatments (0YR, 2YR, and 5YR).

The aforementioned results can have relevant implications in arid and semi-arid regions where water is limited for the majority of irrigation farms. In this context, water use efficiency is an important issue to be considered because with its increase more crops can be irrigated. Based on our outcomes, for a fixed irrigation pattern able to produce approximately immediate soil surface ponding, the use of TWW reduces gradually in time the amount of water entering the soil and increases runoff. To better specify this element, Figure 4 shows the reduction in time of water amount that the soil can absorb in 2YR and 5YR plots with respect to the control plot (0YR) for irrigation duration up to 3 h. From this figure it can be observed that for an irrigation period of 1.5 h the water amount absorbed by the soil in 2YR and 5YR plots is reduced by 41 mm and 47 mm, respectively; for an irrigation stage of 3 h, the infiltrated water depth decreases by 58 and 68 mm, respectively. The water amount that the soil is not able to absorb becomes surface runoff. This implies that for an irrigation scheduled time a reduced water amount is requested in these two plots to avoid an increased runoff. The reduction of absorbable water is here interpreted as gained water and expressed in percentage, with respect to the water infiltrated in the control plot, is defined as irrigation efficiency. The trend of this quantity in the function of irrigation duration is represented in Figure 5, which highlights a decreasing advantage with increasing irrigation periods. For durations ranging from 30 min to 180 min, the irrigation efficiency tends to decrease from 50% to 44% and from 48% to 38% for 5YR and 2YR plots, respectively. In any case, the above values show how after just 2 years the efficiency is significantly increased and after a further 3 years, it becomes almost 50%.

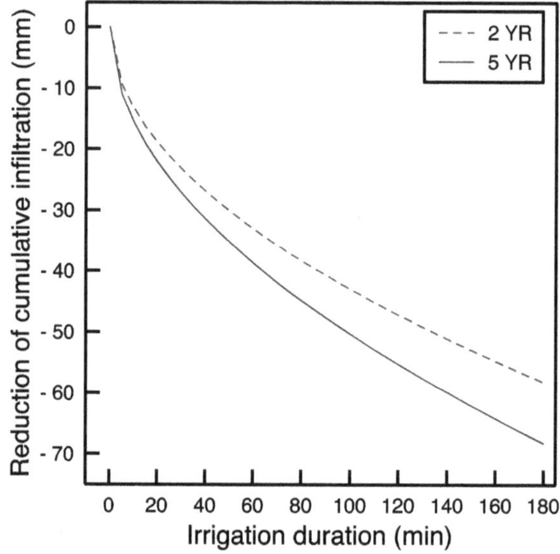

Figure 4. Reduction of water amount absorbable for periods of irrigation up to 3 h in 2YR and 5YR plots if compared with the benchmark plot (0YR).

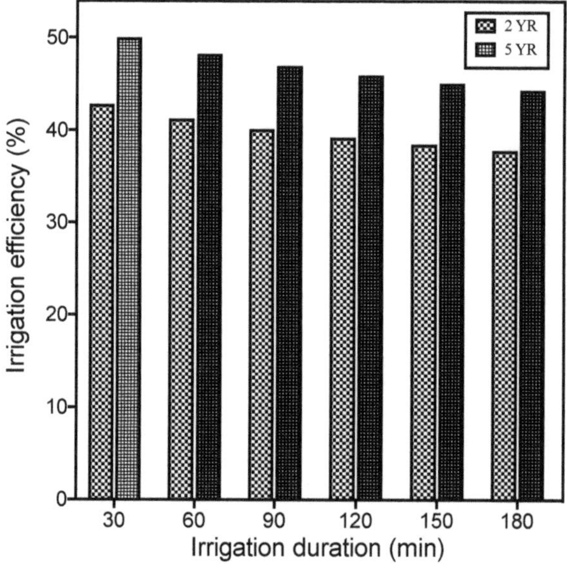

Figure 5. Irrigation efficiency obtained in 2YR and 5YR plots for periods of irrigation up to 3 h in terms of cumulative infiltration reduction if compared with the benchmark plot (0YR).

5. Conclusions

Insights about the effects on the infiltration process produced by continuous use of TWW for irrigation have been provided. Infiltration measurements earlier realized by the Hood infiltrometer in three adjacent field plots characterized by the same soil type and different durations of TWW irrigation (0, 2, and 5 years, named 0YR, 2YR, and 5YR) have been exploited to estimate the associated Philip model parameters, saturated hydraulic conductivity K_s, and sorptivity S. From infiltration simulations performed by the model under the hypothesis of ponded conditions, applicable during irrigation, a quantitative estimate of TWW usage effects has been carried out.

Specifically:

1. The continuous usage of TWW for irrigation determines a lower capacity of water drainage in unsaturated conditions mainly due to the clogging process of the smaller pores by the accumulation of suspended sediments. This leads to a significant decrease of the S parameter in 2YR and 5YR plots, while the saturated hydraulic conductivity linked with the connectivity of larger pores is only weakly affected.

2. The simulations by the Philip model with the decreased values of sorptivity have highlighted reductions of cumulative infiltration in a plot with TWW treatment. For an irrigation pattern with a duration of 1.5 h, the reduction of absorbable water amount with respect to 0YR plot has been found equal to 40% and 47% in 2YR and 5YR plots, respectively, with irrigation duration equal to 3 h the percentages reduce to 38% and 44%, respectively. Equivalent increases of runoff have to be expected if the irrigation water amount remains the same. Hence, the percentage reductions of cumulative infiltration have been here interpreted as water amounts that can be saved for that planned irrigation pattern and have been considered a measure of the gained irrigation efficiency.

3. The above-defined advantages of TWW usage (in terms of irrigation efficiency) slightly decrease with increasing irrigation duration (30 min up to 180 min) ranging from 50% to 44% and from 48% to 38% for the 5YR and 2YR plots, respectively. Anyway, the irrigation efficiency is significant and can be relevant in arid and semi-arid areas.

The results of this research would indicate that the use of treated wastewater leads to the reduction of the required water used to irrigate soil limiting runoff. This is an important outcome of this work that indicates the use of treated wastewater in irrigation practice as one of the possible strategies to be adopted in arid zones to increase irrigation efficiency. However, further investigations are required to establish the link between irrigation efficiency defined here and crop production. Furthermore, complete knowledge of TWW long-term effects on quality characteristics of water and agricultural soils is still lacking.

Author Contributions: Conceptualization, investigation, methodology, writing—original draft preparation and writing—review and editing, A.A.A., M.A.G., M.Z.A., R.M., C.S., T.A.G., and A.F. All authors have read and agreed to the published version of the manuscript.

Funding: This research was financed by the Deanship of Research at the Jordan University of Science and Technology for financial support.

Acknowledgments: The authors are thankful to C.Corradini for proofreading the paper.

References

1. Ravi, V.; Williams, J.R. *Estimation of Infiltration Rate in the Vadose Zone: Compilation of Simple Mathematical Models - Volume I*; US Environmental Protection Agency: Washington, DC, USA, 1998.
2. Machiwal, D.; Jha, M.K.; Mal, B.C. Modelling Infiltration and quantifying Spatial Soil Variability in a Wasteland of Kharagpur, India. *Biosyst. Eng.* **2006**, *95*, 569–582. [CrossRef]
3. Walker, W.R.; Prestwich, C.; Spofford, T. Development of the revised USDA–NRCS intake families for surface irrigation. *Agric. Water Manag.* **2006**, *85*, 157–164. [CrossRef]
4. Gharaibeh, M.A.; Eltaif, N.I.; Al-Abdullah, B. Impact of Field Application of Treated Wastewater on Hydraulic Properties of Vertisols. *Water. Air. Soil Pollut.* **2007**, *184*, 347–353. [CrossRef]
5. Mohammad, M.J.; Mazahreh, N. Changes in Soil Fertility Parameters in Response to Irrigation of Forage Crops with Secondary Treated Wastewater. *Commun. Soil Sci. Plant Anal.* **2003**, *34*, 1281–1294. [CrossRef]
6. Bedbabis, S.; Ben Rouina, B.; Boukhris, M.; Ferrara, G. Effect of irrigation with treated wastewater on soil chemical properties and infiltration rate. *J. Environ. Manage.* **2014**, *133*, 45–50. [CrossRef] [PubMed]

7. Tunc, T.; Sahin, U. The changes in the physical and hydraulic properties of a loamy soil under irrigation with simpler-reclaimed wastewaters. *Agric. Water Manag.* **2015**, *158*, 213–224. [CrossRef]

8. Alizadeh, A.; Bazari, M.E.; Velayati, S.; Hasheminia, M.; Yaghmai, A. Using reclaimed municipal wastewater for irrigation of corn. In *International Workshop on Wastewater Reuse Management*; ICID-CIID: Seoul, Korea, 2011; pp. 147–154.

9. Gharaibeh, M.A.; Ghezzehei, T.A.; Albalasmeh, A.A.; Alghzawi, M.Z. Alteration of physical and chemical characteristics of clayey soils by irrigation with treated waste water. *Geoderma* **2016**, *276*, 33–40. [CrossRef]

10. Schwärzel, K.; Punzel, J. Hood infiltrometer—A new type of tension infiltrometer. *Soil Sci. Soc. Am. J.* **2007**, *71*, 1438–1447.

11. Wooding, R.A. Steady Infiltration from a Shallow Circular Pond. *Water Resour. Res.* **1968**, *4*, 1259–1273. [CrossRef]

12. Gardner, W.R. Some steady-state solutions of the unsaturated moisture flow equation with application to evaporation from a water table. *Soil Sci.* **1958**, *85*, 228–232. [CrossRef]

13. Bardhan, G.; Russo, D.; Goldstein, D.; Levy, G.J. Changes in the hydraulic properties of a clay soil under long-term irrigation with treated wastewater. *Geoderma* **2016**, *264*, 1–9. [CrossRef]

14. Raats, P.A.C. Analytical Solutions of a Simplified Flow Equation. *Trans. ASAE* **1976**, *19*, 683–689. [CrossRef]

15. Philip, J.R. The theory of infiltration: 2. the profile of infinity. *Soil Sci.* **1957**, *83*, 435–448. [CrossRef]

16. Richards, L.A. Capillary conduction of liquids through porous mediums. *J. Appl. Phys.* **1931**, *1*, 318–333. [CrossRef]

17. White, I.; Sully, M.J. Macroscopic and microscopic capillary length and time scales from field infiltration. *Water Resour. Res.* **1987**, *23*, 1514–1522. [CrossRef]

18. Reynolds, W.D.; Topp, G.C. Soil Water Analyses: Principles and Parameters. In *Soil Sampling and Methods of Analysis*; Carter, M.R., Gregorich, G.E., Eds.; CRC Press Taylor & Francis Group: Boca Raton, FL, USA, 2007; pp. 913–937.

19. Viviani, G.; Iovino, M. Wastewater Reuse Effects on Soil Hydraulic Conductivity. *J. Irrig. Drain. Eng.* **2004**, *130*, 476–484. [CrossRef]

Article

Sensibility Analysis of the Hydraulic Conductivity Anisotropy on Seepage and Stability of Sandy and Clayey Slope

Shuyang Yu [1], Xuhua Ren [1,2,*], Jixun Zhang [1], Haijun Wang [3] and Zhitao Zhang [3]

[1] College of Water Conservancy and Hydro-Power Engineering, Hohai University, Nanjing 210098, China; yushuyang_hhu@163.com (S.Y.); zhangjixun@hhu.edu.cn (J.Z.)
[2] Collaborative Innovation Center on Water Safety and Water Science, Hohai University, Nanjing 210098, China
[3] State Key Laboratory of Hydrology-Water Resource and Hydraulic Engineering, Nanjing Hydraulic Research Institute, Nanjing 210029, China; wanghaijun@163.com (H.W.); zhangzhitao1234@sina.com.cn (Z.Z.)
*   Correspondence: renxh@hhu.edu.cn; Tel.: +86-133-0518-1963

Received: 1 December 2019; Accepted: 16 January 2020; Published: 18 January 2020

Abstract: Evaluation of slope stability under rainfall is an important topic of Geotechnical Engineering. In order to study the influence of anisotropy ratio ($k_r = k_x/k_y$) and anisotropy direction (α) on the seepage and stability of a slope, the SEEP/W and SLOPE/W modules in Geo-studio were utilized to carry out the numerical analysis of a homogeneous slope in Luogang District, Guangzhou City, China, which is based on the theory of unsaturated seepage and stability. Two kinds of soils (clay and sand) were included. Results show that: For sandy soil slope, the increase of k_r promotes the rainfall infiltration, and the decrease of α prevents the rainfall infiltration. The maximum water content of the surface (MWCS) reaches maximum with the increase of k_r and α. The rising height of groundwater (RHG) is −3–4 m and the safety factor (SF) is 1.3–1.7. For clayey soil slope, variations of k_r and α have little impact on the seepage characteristics and slope stability. The MWCS remains almost the same. The rainfall infiltration depth (RID) is 0.5–1 m and the SF is about 1.7. Therefore, for sandy soil slope, it is not only necessary to consider the influence of k_r, but also the influence of α. For clayey soil slope, it can be treated as isotropic material to simplify calculation.

Keywords: hydraulic conductivity anisotropy; sandy soil slope; clayey soil slope; seepage characteristics; slope stability; numerical simulation

---

1. Introduction

Slope stability is an important engineering problem in the geotechnical field, for example, the slope stability of the excavation of a foundation pit [1], the regulation of a riparian slope [2], and the stability of a high slope under complex geological environment in a reservoir area [3], etc. Slope failures mainly include external and internal causes. The external causes include rainfall [4–6], reservoir water level fluctuations [7–9], earthquakes [10–12], human activities such as excavation or blasting [1,3–15], etc. The internal causes are mainly affected by the properties of the slope soil, which include soil types [16–18], unsaturated characteristics [19–21], soil strength [22–24], etc. Rainfall is the key factor triggering the landslide, which accounts for 51% of all the landslide disasters, according to some relevant investigations [25]. Rainfall usually occurs in rainy seasons, which is concentrated and has a long duration and leads to landslide easily. The main reasons for slope instability caused by rainfall are as follows: (1) Rainfall increases the groundwater level inside the slope, which reduces the effective stress and shear strength of the soil. (2) Rainfall increases the slide force of the slope, which aggravates the slope instability. The main consequence for slope instability caused by rainfall is reflected in these two aspects: (1) Landslide causes damage to buildings in the area where the slope is located. (2) Landslide threatens the lives and properties of the residents around the disaster area. For example, on 16 September 2011, heavy rainfall (250 mm/day, cumulative 430 mm) in Nanjiang County, Sichuan Province, China, induced thousands of landslides, and on 29 April 2015, the groundwater level in Dangchuan, Heifangtai, Gansu Province, China, rose due to heavy rainfall, which caused a large-scale slope failure [26]. Therefore, it is important to grasp the law and influencing factors of slope instability caused by rainfall in order to correctly understand the mechanism of rainfall infiltration and to prevent and control landslide disasters.

Scholars have conducted large numbers of research studies on the rainfall infiltration mechanism. The research results mainly focused on theory, experiments, and numerical simulations. With respect to theoretical research, Green-Ampt Semi-Analytical Method [27] was the first method to describe the transient infiltration process of rain water in unsaturated soils. It is assumed that the wetting front moves down along the depth direction and the velocity remains unchanged. The volume of water content of the soil after the wetting front is θ_0 (completely saturated), and the initial water content of the soil before the front is θ_i. The rainfall intensity is always higher than the infiltration capacity of the soil, and the actual infiltration rate is equal to the infiltration capacity of the soil. Mein et al. [28] improved the Green-Ampt model and rainfall infiltration process was divided into two periods. The first stage is the free infiltration period, and the rainfall intensity in this stage is less than the soil infiltration capacity. The second stage is the ponding infiltration period, and the rainfall intensity is greater than the soil infiltration capacity. Chu et al. [29] divided the non-uniform rainfall process into several uniform periods, and calculated the infiltration process according to whether there will be ponding. Chen et al. [30] derived a uniform slope rainfall infiltration model based on the Green-Ampt model. For experimental research, Wu et al. [31] carried out the laboratory model test of landslide under artificial rainfall, and the influence of rain water infiltration on the slope failure was analyzed. Li et al. [32] studied the influence of rainfall on the internal mechanical response characteristics of slope based on fiber grating monitoring technology. Zhang et al. [33] systematically investigated the stability of Xiakou slope under rainfall based on the field monitoring data. In the aspect of numerical simulation, Hao et al. [34] simulated the variations of stability of a typical slope under rainfall based on the limit equilibrium method. Wang et al. [35] used XFEM (extended finite element method) to simulate the crack propagation process in a slope under heavy rain. Dou et al. [36] considered the spatial variability of hydraulic conductivity of slope soil based on the Monte Carlo method, and simulated the seepage characteristics and slope stability. However, most of the previous studies regarded the slope materials as isotropic materials. According to the SEM (scanning electron microscope) microcosmic study of Song et al. [37], the anisotropy of permeability coefficient is caused by the flocculation microstructure of clay and other soils, meanwhile, soil permeability coefficient anisotropy is greatly affected by dry density and freeze-thaw cycles [38]. Generally speaking, the hydraulic conductivity anisotropy ratio (k_x/k_y) can reach 2–10, and it is possible to reach 100 for clayey soil [39]. There exist joint cracks in rock slope [40,41], which lead to strong anisotropy of seepage characteristics. The coefficient anisotropy not only has great influences on the transient seepage but also has an impact on the safety factor of the slope. According to Mahmood et al. [42], the difference of slope safety factors between considering and not considering soil anisotropy will be about 40% [43]. But most of the previous studies ignored the seepage anisotropy of slope soil, and the research results of soil anisotropy of slope were few and incomplete. Yeh et al. [44] took into account the hydraulic conductivity anisotropy ratio of soil slope, and simulated the seepage characteristics and local safety factors, but ignored the hydraulic conductivity anisotropy direction. In fact, the horizontal permeability coefficient k_x and vertical permeability coefficient k_y coincide with the natural coordinate axis only in some special cases such as layered crushed earth dam or naturally deposited layered soil. In nature, there are more cases where the anisotropy principal direction does not coincide with the coordinate axis. The actual conditions cannot be accurately reflected only by considering the hydraulic conductivity anisotropy ratio.

In view of the shortcomings of previous studies, the mathematical definition of hydraulic conductivity anisotropy ratio and direction are firstly described in this paper. The SEEP/W and SLOPE/W modules in Geo-studio were utilized to carry out the numerical analysis of a homogeneous slope in Luogang District, Guangzhou City, China [45]. Geo-studio is a professional software suitable for analyzing the seepage and stability of soil slopes, and the numerical results were consistent with the experimental results and field investigations. For example, Jiang et al. [46] analyzed the seepage and stability of a cracked slope with SEEP/W and SLOPE/W. Duong T.T. [2] used the Geo-studio program to study the effects of soil hydraulic conductivity and rainfall intensity on riverbank stability. Muqdad Al-Juboori et al. [47–49] conducted the machine learning research based on SEEP/W. Therefore, this software was utilized in this paper to analyze the effect of hydraulic conductivity anisotropy. Two kinds of soil (clay and sand) and the hydraulic conductivity anisotropy ratio and direction were included. Then the volume water content, rainfall infiltration depth (RID), rising height of groundwater (RHG) and maximum water content of the surface (MWCS) of three typical sections of slope (top, middle, and toe) were analyzed in detail. Finally, the safety factors (SF) of the slope under different conditions were evaluated. The research results provide some references for the understanding of the seepage anisotropy law and prevention of landslides.

2. Methods and Theory

2.1. Theory of Unsaturated Seepage

The SEEP/W module in Geo-studio was utilized to simulate the rainfall infiltration, and the seepage control equation in the SEEP/W was derived from the saturated and unsaturated Darcy's law [50], which can be expressed as

$$\frac{\partial}{\partial x}\left(k_x \frac{\partial H}{\partial x}\right) + \frac{\partial}{\partial y}\left(k_y \frac{\partial H}{\partial y}\right) + Q = m_w \gamma_w \frac{\partial H}{\partial t}. \tag{1}$$

In Equation (1), x and y are the coordinates in the direction of x and y, k_x is the hydraulic conductivity in the x direction, k_y is the hydraulic conductivity in the y direction, H is the total head, Q is the applied boundary flux, t is the time, m_w is the slope of the storage curve, and γ_w is the unit weight of water.

Applying the Galerkin method of weighed residual to the governing differential equation, the finite element for two-dimensional seepage equation can be derived as

$$\tau \int_A ([B]^T [C][B])dA\{H\} + \tau \int_A (\lambda\langle N\rangle^T \langle N\rangle)dA\{H\}, t = q\tau \int_L (\langle N\rangle^T)dL. \tag{2}$$

In Equation (2), $[B]$ is the gradient matrix, $[C]$ is the element hydraulic conductivity matrix, $[H]$ is the vector of nodal heads, $<N>$ is the vector of interpolating function, q is the unit flux across the edge of an element, τ is the thickness of an element, λ is the storage term for a transient seepage equal to $m_w \gamma_w$, A is a designation for summation over the area of an element, and L is a designation for summation over the edge of an element.

In an abbreviated form, the finite element seepage equation can be expressed as

$$[K]\{H\} + [M]\{H\}, t = \{Q\}. \tag{3}$$

In Equation (3), $[K]$ is the element characteristic matrix, $[M]$ is the element mass matrix, and $\{Q\}$ is the element applied flux vector.

2.2. Effect of Negative Pore-Water Pressures

In locations above the groundwater table, the pore-water pressure in the soil is negative relative to the pore-air pressure. This negative pore-water pressure is commonly referred to as the matric suction of the soil. Under negative pore-water pressure conditions the shear strength may not change at the same rate as for total and positive pore-water pressure changes. Therefore, a modified form of the Mohr–Coulomb equation must be used to describe the shear strength of an unsaturated soil (i.e., the soil with negative pore-water pressures). The shear strength equation is [51]

$$s = c' + \sigma_n \tan \varphi' + (u_a - u_w) \tan \varphi_b. \tag{4}$$

In Equation (4), s is the unsaturated shear strength, c' is the cohesive strength, φ' is the frictional strength, φ_b is an angle defining the increase in shear strength for an increase in suction, u_a is the pore-air pressure, and u_w is the pore-water pressure.

2.3. Safety Factor for Unsaturated Soil

SLOPE/W adopts the Morgenstern–Price method based on limit equilibrium theory to calculate the safety factor. The modified method strictly satisfies the force balance and torque balance, and the calculation accuracy is high. The expression is listed below:

$$F_s = \frac{\sum\limits_{i=1}^{n_s} \dfrac{c'_i b_i + (W_i + P_i \cos \beta_i - u_a b_i) \tan \varphi'_i + (u_a - u_w) b_i \tan \varphi_b}{\left[1 + (\tan \varphi'_i \tan \alpha_i)/F_s\right] \cos \alpha_i}}{\sum\limits_{i=1}^{n_s} W_i \sin \alpha_i - r_i P_i}. \tag{5}$$

In Equation (5), c_i' is the cohesive strength for every soil slice, i is the soil slice number, W_i is the weight of every soil slice, P_i is the water pressure, β_i is the angle of the bottom of the soil slice, b_i is the length of every soil slice, φ_i' is the frictional strength for every soil slice, r_i is the radius of the sliding arc, and F_s is the safety factor.

3. Numerical Model Framework

3.1. Numerical Model and Boundary Conditions

The case study is a homogeneous slope in Luogang District, Guangzhou City, China [46]. The slope height is 16 m, which is divided into 2 grades with a width of 2 m. In order to reduce the influence of boundary conditions, the range was extended and the model was divided into 13,484 nodes and 13,279 units, which are shown in Figure 1.

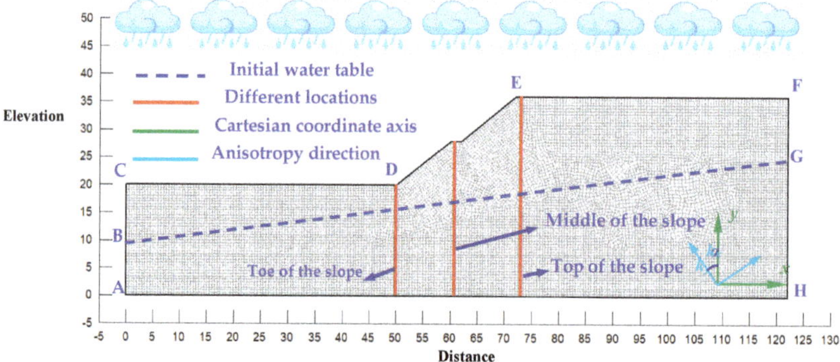

Figure 1. Illustration of slope model.

To investigate the seepage characteristics at different positions, three sections were set to reflect the effect of hydraulic conductivity anisotropy, whose positions were $x = 73$ m (top of the slope), $x = 61$ m (middle of the slope), and $x = 50$ m (toe of the slope). The boundary conditions were as follows: AB and GH were the fixed water level boundaries of 9 m and 24 m, respectively. CDEF was the rainfall infiltration boundary. BC and GF were small flux boundaries. AH was the impervious boundary.

3.2. Unsaturated Soil Properties

The soil-water characteristic curves (SWCC) adopted the Fredlund and Xing model, which can be written as [52]

$$k_w = k_s \frac{\sum\limits_{i=j}^{N} \frac{\Theta(e^y)-\Theta(\Psi)}{e^{y_i}} \Theta'(e^{y_i})}{\sum\limits_{i=1}^{N} \frac{\Theta(e^y)-\Theta_s}{e^{y_i}} \Theta'(e^{y_i})}. \tag{6}$$

In Equation (6), k_w is the calculated conductivity for a specified water content or negative pore-water pressure, k_s is the measured saturated conductivity, Θ_s is the volumetric water content, e is the natural number 2.71828, y is a dummy variable of integration representing the logarithm of negative pore-water pressure, i is the interval between the range of j to N, j is the least negative pore-water pressure to be described by the final function, N is the maximum negative pore-water pressure to be described by the final function, Ψ is the suction corresponding to the jth interval, Θ' is the first derivative of the equation, and Θ can be described as

$$\Theta = C(\Psi) \frac{\Theta_s}{\left\{\ln\left[e + \left(\frac{\Psi}{a}\right)^n\right]\right\}^m}. \tag{7}$$

In Equation (7), a is the air-entry value of the soil, n is a parameter that controls the slope at the inflection point in the volumetric water content function, m is a parameter that is related to the retention capacity, and $C(\Psi)$ is a correcting function defined as

$$C(\Psi) = 1 - \frac{\ln(1 + \frac{\Psi}{C_r})}{\ln(1 + \frac{1000000}{C_r})}. \tag{8}$$

In Equation (8), C_r is a constant related to the matric suction corresponding to the retention capacity.

The sandy soil and clayey soil were selected for analysis, which represent the high and low permeability [53], as shown in Figure 2. The unsaturated parameter values are shown in Table 1 [54], and the SWCC curves are shown in Figure 3.

Figure 2. Typical appearance of sandy and clayey soil. (**a**) Clayey soil. (**b**) Sandy soil.

Table 1. Unsaturated parameter values.

Soil Type	SWCC Parameters				Permeability Coefficient	
	a/kPa	*m*	*n*	*θ*/%	k_x/(m/s)	k_x/(m/day)
Sand	10	1	1	45	10^{-4}	8.64
Clay	100	1	1	45	10^{-6}	0.0864

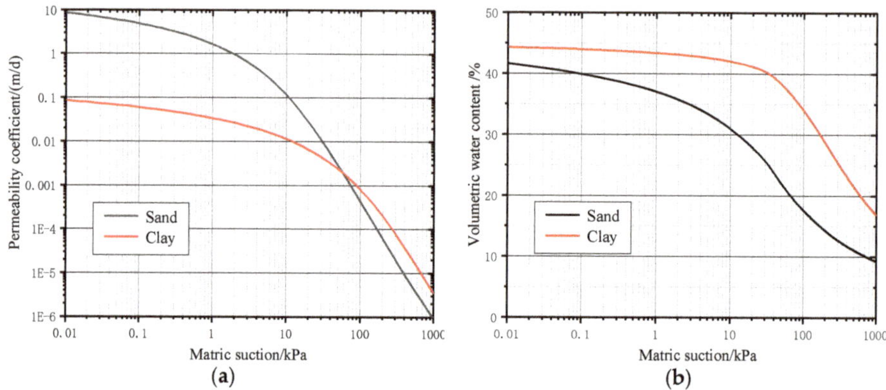

Figure 3. SWCC curves. (**a**) Permeability coefficient function. (**b**) Volume water content function.

3.3. Definition of Anisotropy and Calculation Conditions

Previous studies mostly ignored the anisotropy ratio and direction. In fact, anisotropy widely exists in the soil. For the hydraulic conductivity matrix *[C]* in Equation (2), it can be expressed as

$$[C] = \begin{bmatrix} C_{11} & C_{12} \\ C_{21} & C_{22} \end{bmatrix}. \tag{9}$$

In Equation (9), $C_{11} = k_x\cos^2\alpha + k_y\sin^2\alpha$, $C_{22} = k_x\sin^2\alpha + k_y\cos^2\alpha$, and $C_{12} = C_{21} = k_x\sin\alpha\cos\alpha + k_y\sin\alpha\cos\alpha$. The k_x, k_y, the anisotropy direction α can be defined according to Figure 2. The k_x is the horizontal permeability coefficient, k_y is the vertical permeability coefficient, and α is the direction between k_y and y axis. When $\alpha = 0°$, *[C]* is reduced to

$$[C] = \begin{bmatrix} k_x & 0 \\ 0 & k_y \end{bmatrix}. \tag{10}$$

Equation (10) was adopted in [44], with only considering the anisotropy ratio $k_r = k_x/k_y$. However, the definition of anisotropy not only includes the ratio k_r but also the direction α. Previous investigations ignored the anisotropy, especially the anisotropy direction α.

To completely discuss the anisotropy of sandy soil and clayey soil, including the anisotropy ratio k_r and the anisotropy direction α, the calculation conditions are shown in Table 2, which includes the anisotropy ratio $k_r = 1, 10, 50, 100$, and the anisotropy direction $\alpha = 0°, 15°, 30°, 45°, 60°, 75°$, and $90°$. The range of k_r and α were selected according to Gilbert el al. [39]. To reflect the influence of the heavy rainfall, the rainfall intensity is set to 10^{-6} m/s, and the rainfall duration time is set to 120 h. Meanwhile, 120 h of rainfall stop was also considered.

Table 2. Different calculation conditions.

Rainfall Intensity (m/s)	Soil Type	k_x (m/s)	k_r	$\alpha/°$
10^{-6}	Sand	10^{-4}	$\begin{bmatrix} 1 \\ 10 \\ 50 \\ 100 \end{bmatrix}$	$\begin{bmatrix} 0 \\ 15 \\ 30 \\ 45 \\ 60 \\ 75 \\ 90 \end{bmatrix}$
	Clay	10^{-6}	$\begin{bmatrix} 1 \\ 10 \\ 50 \\ 100 \end{bmatrix}$	$\begin{bmatrix} 0 \\ 15 \\ 30 \\ 45 \\ 60 \\ 75 \\ 90 \end{bmatrix}$

4. Results and Discussions

4.1. Initial Conditions

Initial conditions are important for further numerical simulations. In order to determine the initial conditions more accurately in this paper, the maximum negative pore water pressure of −25 kPa, −50 kPa, and −75 kPa and the specified annual average rainfall infiltration were numerically simulated, and the pore pressure variation of sandy soil and clayey soil are shown in Figure 4. Under the annual average rainfall infiltration, the initial pore pressure of sand and clay was obviously different in that the initial pore pressure of sandy soil slope was slightly larger than clayey soil slope, but the distribution along the elevation was similar, which was reflected in that the initial pore water pressure firstly remained unchanged then gradually increased along the elevation. The maximum negative pore-water pressure was close to −50 kPa, which was selected as the initial condition of all the calculation conditions in this paper.

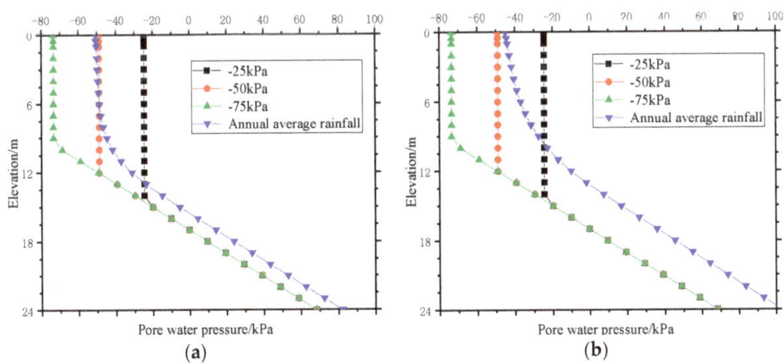

Figure 4. Initial pore pressure distribution. (**a**) Sandy soil. (**b**) Clayey soil.

4.2. Variation of Volumetric Water Content

According to the calculation conditions in Table 2, we carried out a total number of 54 numerical simulations, and obtained, in all, 162 sections of volumetric water content variation. For ease of reading, this section will carry out the discussion based on the classification of sandy soil and clayey soil slope. The variations of volumetric water content of different α values with $k_r = 10$ and $k_r = 100$ are shown in Figures 5 and 6 to illustrate the influence of anisotropy direction α on seepage characteristics, and different k_r values with $\alpha = 0°$, $45°$, and $90°$ are also displayed in Figures 7 and 8 to show the impact of anisotropy ratio k_r. The volumetric water content of the 120th h is only shown in Figures 5–8.

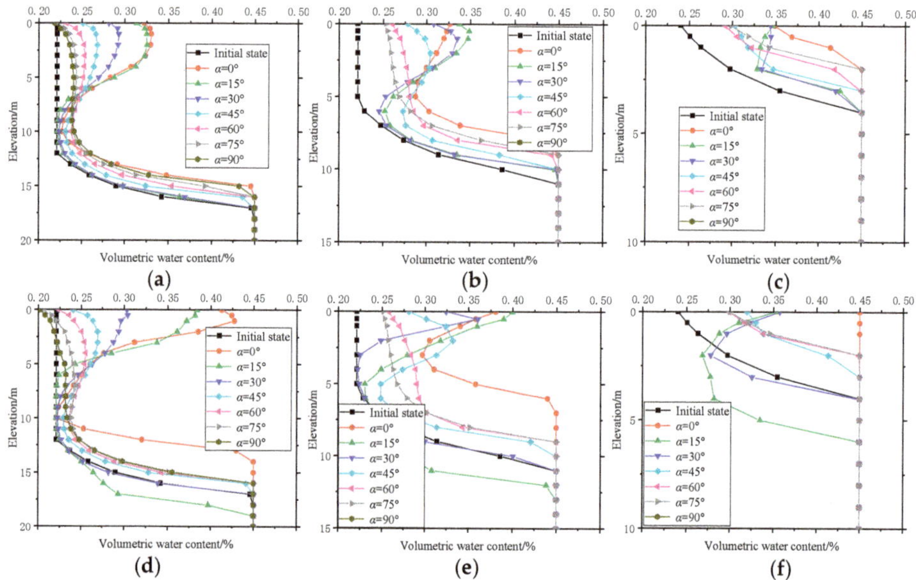

Figure 5. Variation of volumetric water content under different α values for sandy soil. (**a**) Top of the slope with $k_r = 10$. (**b**) Middle of the slope with $k_r = 10$. (**c**) Toe of the slope with $k_r = 10$. (**d**) Top of the slope with $k_r = 100$. (**e**) Middle of the slope with $k_r = 100$. (**f**) Toe of the slope with $k_r = 100$.

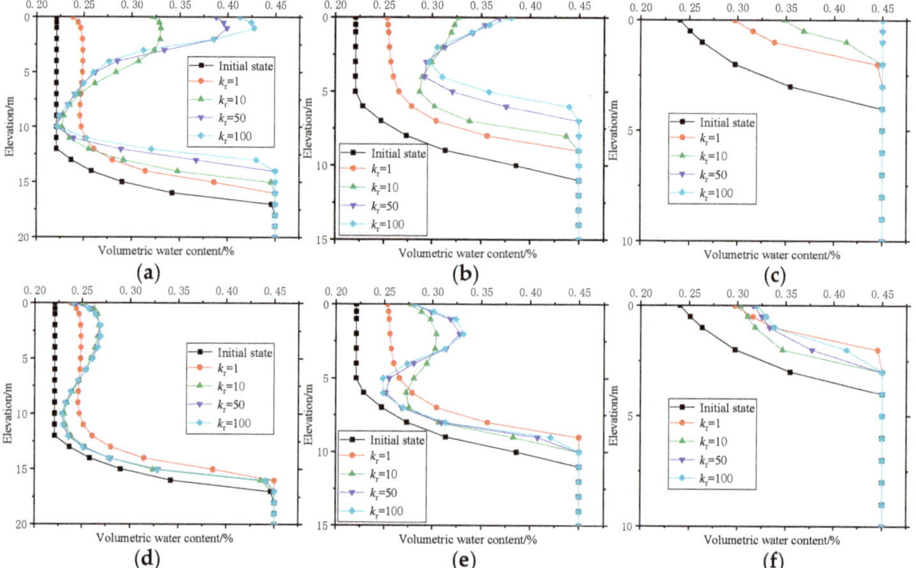

Figure 6. *Cont.*

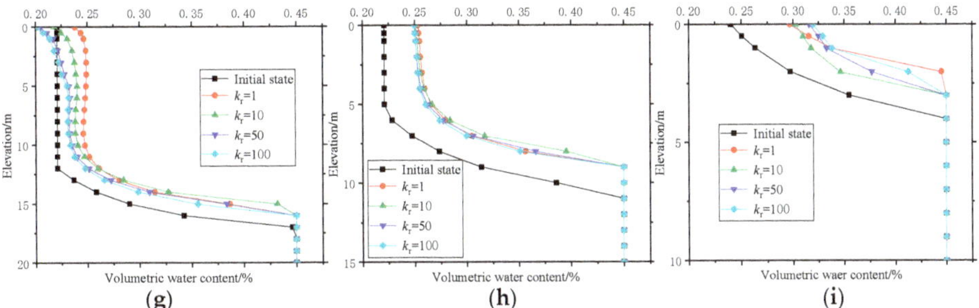

Figure 6. Variation of volumetric water content under different k_r values for sandy soil. (**a**) Top of the slope with $\alpha = 0°$. (**b**) Middle of the slope with $\alpha = 0°$. (**c**) Toe of the slope with $\alpha = 0°$. (**d**) Top of the slope with $\alpha = 45°$. (**e**) Middle of the slope with $\alpha = 45°$. (**f**) Toe of the slope with $\alpha = 45°$. (**g**) Top of the slope with $\alpha = 90°$. (**h**) Middle of the slope with $\alpha = 90°$. (**i**) Toe of the slope with $\alpha = 90°$.

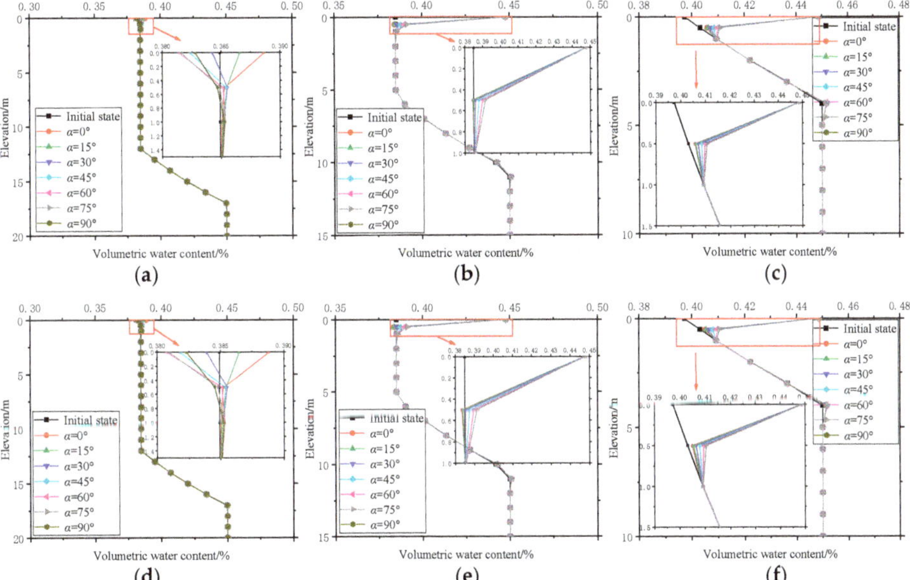

Figure 7. Variation of volumetric water content under different α values for clayey soil. (**a**) Top of the slope with $k_r = 10$. (**b**) Middle of the slope with $k_r = 10$. (**c**) Toe of the slope with $k_r = 10$. (**d**) Top of the slope with $k_r = 100$. (**e**) Middle of the slope with $k_r = 100$. (**f**) Toe of the slope with $k_r = 100$.

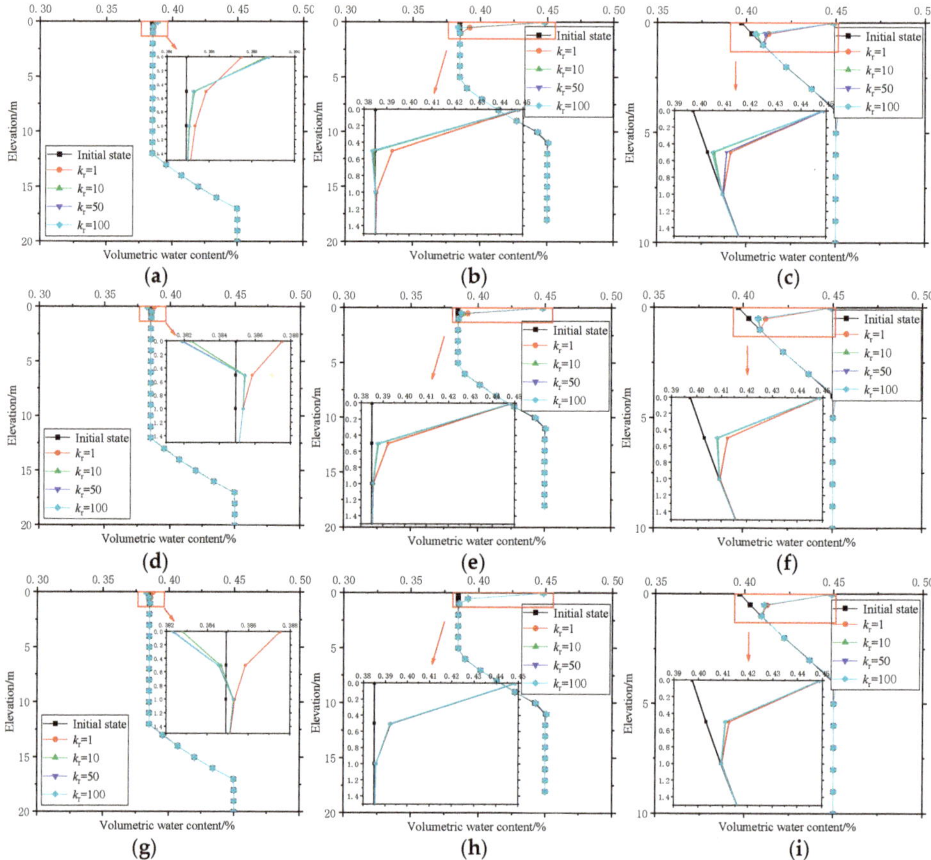

Figure 8. Variation of volumetric water content under different k_r values for clayey soil. (**a**) Top of the slope with $\alpha = 0°$. (**b**) Middle of the slope with $\alpha = 0°$. (**c**) Toe of the slope with $\alpha = 0°$. (**d**) Top of the slope with $\alpha = 45°$. (**e**) Middle of the slope with $\alpha = 45°$. (**f**) Toe of the slope with $\alpha = 45°$. (**g**) Top of the slope with $\alpha = 90°$. (**h**) Middle of the slope with $\alpha = 90°$. (**i**) Toe of the slope with $\alpha = 90°$.

4.2.1. Sandy Soil

The variations of volumetric water content of sandy soil for different sections under different anisotropy direction α values are shown in Figure 5.

For the top of slope, the volumetric water content on the surface decreased with the increase of α. This is because k_x was greater than k_y. When $\alpha = 0°$, the vertical permeability reached the minimum, so the rainfall was hard to infiltrate, and the rain water accumulated in the shallow part of the slope. With the increase of α, however, the vertical permeability increased, and the rain water was easier to infiltrate into the deep part, thus leading to the decrease of the surface volumetric water content. What should be noticed is that when the anisotropy ratio was small (i.e., $k_r = 10$), the impact of rainfall on the volumetric water content was mainly reflected on the slope surface; but when the anisotropy ratio was larger (i.e., $k_r = 100$), not only the slope surface but also the deep area were violently influenced, especially for $\alpha = 0°$ and $\alpha = 15°$.

For the middle of slope, the volumetric water content was affected by the combined effect of rainfall infiltration and the rainfall excretion from the slope top. The increase of α also decreased the volumetric water content on the surface as was illustrated in the previous paragraph. The combined effect, however, made the maximum surface volumetric water content happen at $\alpha = 15°$.

For the toe of slope, the variation of volumetric water content was similar to the slope top and middle In the condition of $\alpha = 0°$, however, the whole section did not reach saturation with $k_r = 10$, but reached saturation with $k_r = 100$.

Figure 6 shows the variation of volumetric water content under different k_r values. For the top of slope, the volumetric water content under different k_r values was greatly affected by the values of α. When α was relatively small (i.e., $\alpha = 0°$), the surface volumetric water content increased with the increase of k_r. This is because the increase of k_r decreased the horizontal permeability, thus leading to a higher surface volumetric water content. Meanwhile, rain water was easier to infiltrate through the horizontal direction when k_r decreased, which rose the underground water level, thus leading to the increase of the deep volumetric water content. Yeh H.F. et al. (2018) [44] conducted relevant.simulations and the results were similar to current research results. However, the increase of α values made the difference of volumetric water content between different k_r values smaller. This is because the vertical permeability decreased with the increase of α, and the rain water drained away rather than infiltrated directly into the slope soil.

For the middle of slope, the variation range of volumetric water content was larger than that of the slope top, and under the combined effect, the value of volumetric water content was also larger than that of the slope top.

For the toe of the slope, due to its lower terrain, the height of the initial water table to the surface of the slope toe was smaller, and it was affected not only by rainfall but also by the rain water from the slope middle. So it had a smaller unsaturated area and a larger volumetric water content. What should be noticed is that the whole section reached saturation when $\alpha = 0°$ and $k_r = 50$ and 100, and did not reach saturation in other conditions.

4.2.2. Clayey Soil

The variations of volumetric water content under different α values are shown in Figure 7, and the variation of volumetric water content under different k_r values are shown in Figure 8.

As can be seen in Figures 7 and 8, the differences between different k_r and α values was relatively small for clayey soil. This is because the clayey soil had a relatively lower permeability, and the rainfall was difficult to infiltrate. Only the rainfall infiltration depth varied for different locations of the slope. The rainfall infiltration depth increased with the decrease of distance to the slope toe, and the deep volumetric water content almost did not change.

Figure 6 shows the variations of volumetric water content for sandy slope. The permeability coefficient was high, which resulted in the obvious change in the variation of volumetric water content, and, in fact, the bending reflected the accumulation of rain water on the shallow part of the slope. However, it was different in the clay slope, which is shown in Figure 8. The low permeability coefficient made the rain water difficult to infiltrate into the soil, and the no-ponding boundary allowed the excess water to move away from the boundary. So once the rainfall stopped, the slope surface did not have rainfall infiltration boundaries anymore, and the bending effect was not so obvious. We can see the bending effect in Figure 8b,d, which was located in the slope middle, but the slope top (Figure 8a,c) was regarded as unchanged.

4.3. Analysis of Rainfall Infiltration Depth, Rising Height of Groundwater, and Maximum Water Content of the Surface

As can be inferred from Section 4.2, the hydraulic conductivity anisotropy ratio k_r and α have a great impact on the seepage characteristics of the slope. In order to evaluate this comprehensively, the rainfall infiltration depth (RID), rising height of groundwater (RHG), and the maximum water content of the surface (MWCS) were defined, as shown in Figure 9. Figure 9a shows distribution of volumetric water content during the rainfall process each day at the toe section of clayey slope.

The surface volumetric water content gradually increased until the rain stopped. So the MWCS was defined to illustrate the saturation on the slope surface, which was the maximum surface water content when the rain stopped. During the rainfall, the wetting front developed into the deep area, and the RID is defined to characterize the influence of the rainfall on the slope deep, which was the height of the turning point to the slope surface. For sandy soil, as shown in Figure 9b, not only the surface volumetric water content but also the rising height of groundwater increased. So the RHG was defined to express the impact of rainfall on the sandy slope, which was the height of the turning point of the volumetric water content when the rain stopped to the turning point of the initial volumetric water content. What should be noticed is that there was no noticeable effect for the second to fifth rainfall day on the water content. This is because the hydraulic conductivity anisotropy was $k_r = 10$ and $\alpha = 0$. So for the sandy slope, rain water was hard to spread, and the difference was relatively small, as shown in Figure 10a. For clay soil, due to its low permeability, the difference seemed to be less obvious.

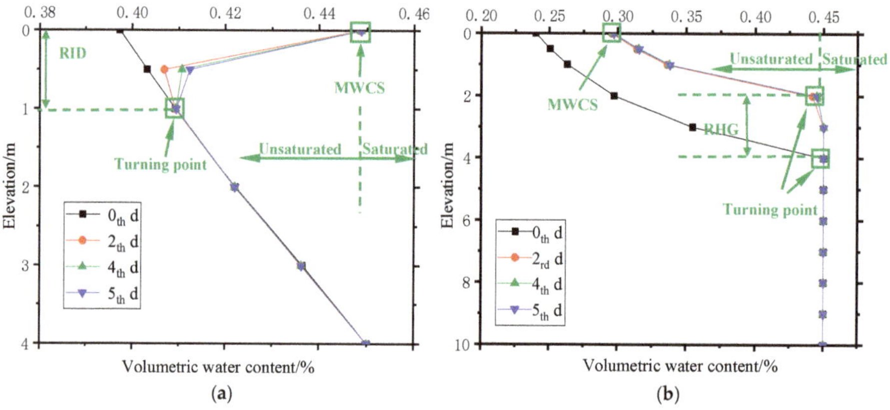

Figure 9. Variation of volumetric water content for sandy and clayey soil. (**a**) Sandy soil. (**b**) Clayey soil.

The MWCS for clayey and sandy soil slope is shown in Figure 10. For sandy soil slope, the MWCS increased to the maximum with the increase of k_r and α. When it came to the slope toe, the value of MWCS became larger than slope top and slope middle, which means that the slope toe was easier to reach saturation during rainfall. For clayey soil slope, the MWCS under different k_r and α values was almost the same for the same section.

Figure 11 shows the variation of RID for clayey soil slope and the variation of RHG for sandy soil slope. For sandy soil slope, rainfall mainly caused the underground water level to rise, and the rising height was −3–4 m. What to be stressed is that when $k_r = 100$, $\alpha = 15°$, the height of groundwater level decreased. We inferred this may be due to the fact that the vertical permeability coefficients were relatively small and the groundwater level was readjusted. For clayey soil slope, the variations of k_r and α had little influence on the RID. For slope top, the RID was 0.5 m, for slope middle it was 0.5–1 m, and for slope toe it was 1 m.

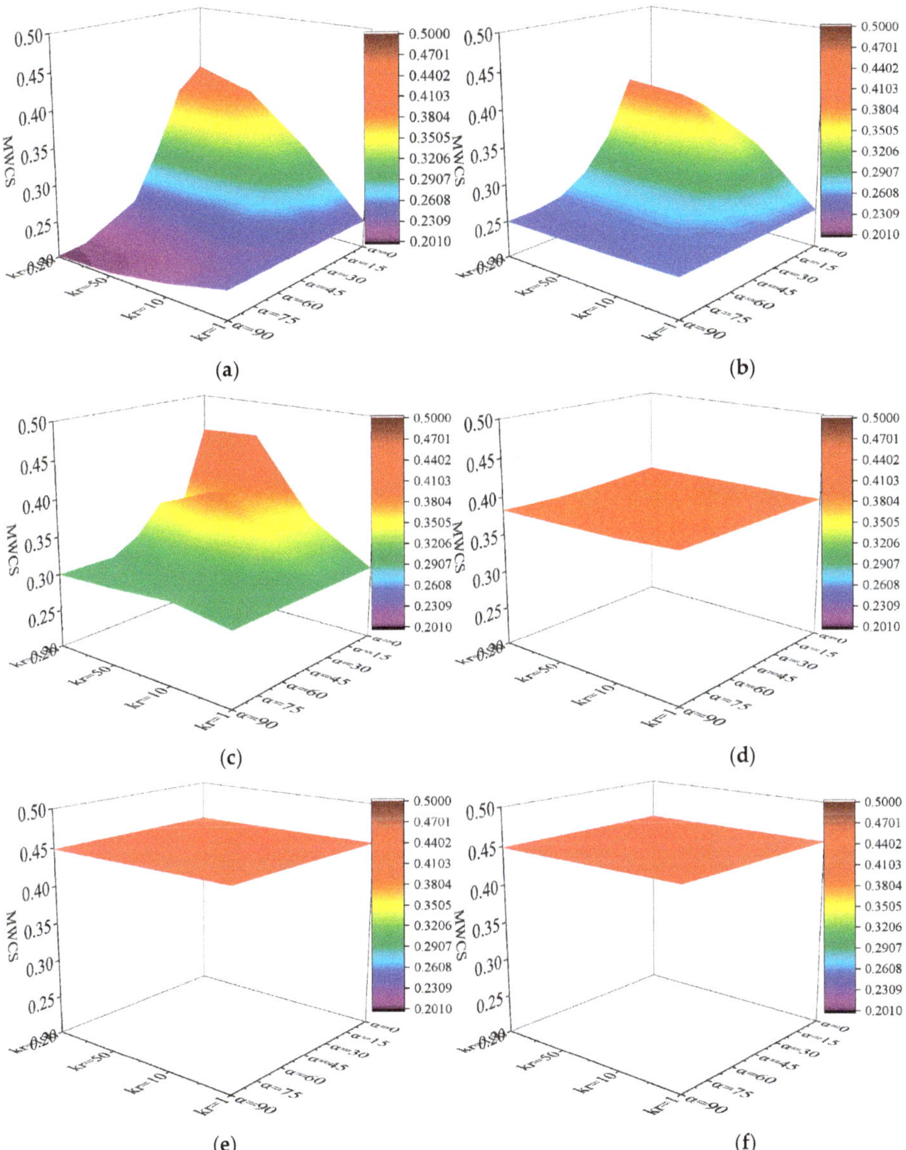

Figure 10. Variation of MWCS for clayey and sandy soil. (**a**) Top of the slope for sandy soil. (**b**) Middle of the slope for sandy soil. (**c**) Toe of the slope for sandy soil. (**d**) Top of the slope for clayey soil. (**e**) Middle of the slope for clayey soil. (**f**) Toe of the slope for clayey soil.

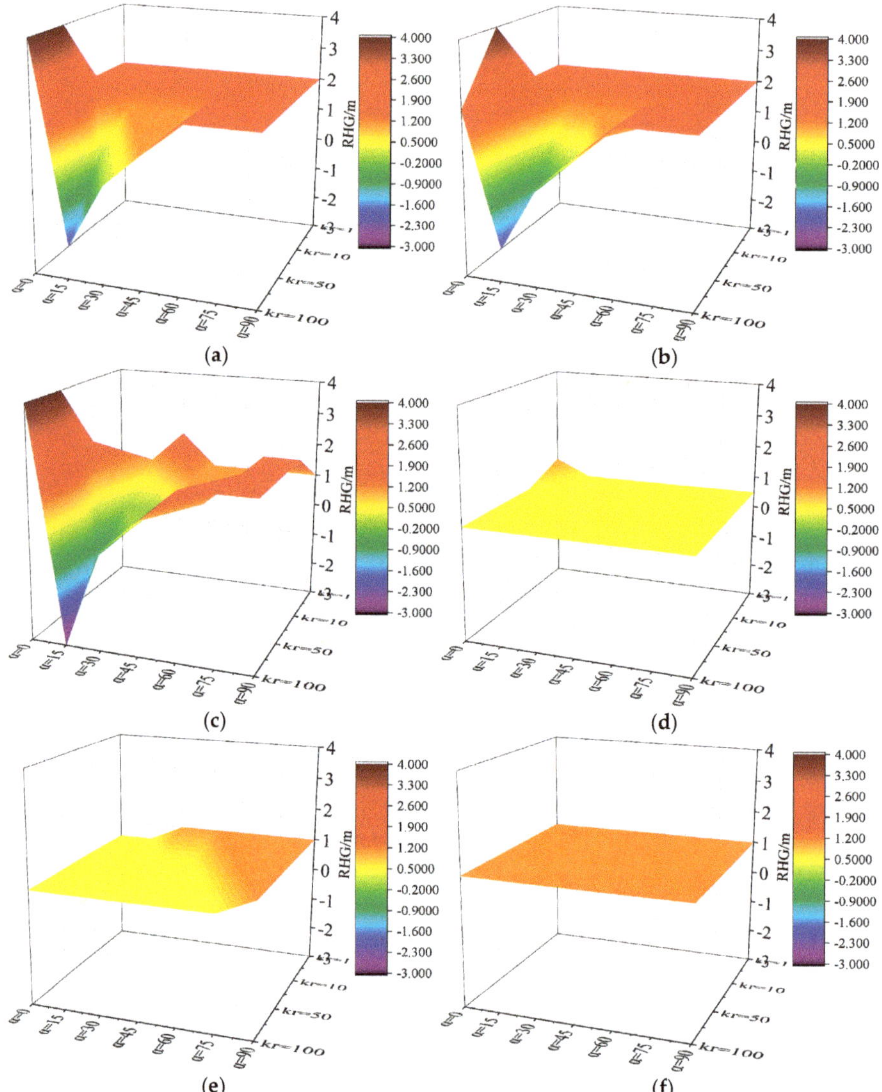

Figure 11. Variation of RID and RHG for clayey and sandy soil. (**a**) Top of the slope for sandy soil. (**b**) Middle of the slope for sandy soil. (**c**) Toe of the slope for sandy soil. (**d**) Top of the slope for clayey soil. (**e**) Middle of the slope for clayey soil. (**f**) Toe of the slope for clayey soil.

4.4. Safety Factors

The SLOPE/W module was utilized in this section to calculate the safety factors based on the Equation (5). In order to control the variables, the calculation parameters of soil strength were set according to Tang et al. [54], where $c' = 10$ kPa, $\varphi' = 26°$, and $\varphi_b = 26$. The variations of safety factors (SF) under different k_r and α values are shown in Figure 12.

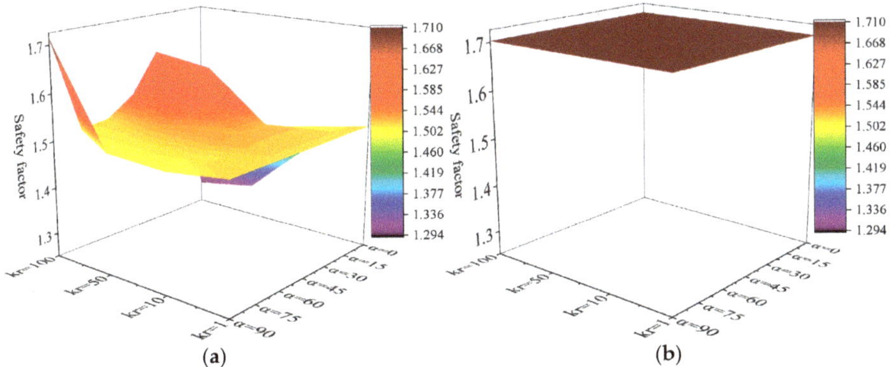

Figure 12. Variation of SF for clayey and sandy soil slope. (**a**) Sandy soil slope. (**b**) Clayey soil slope.

The variations of k_r and α had a great impact on the SF for sandy soil slope. The minimum SF (1.3) happened when $k_r = 100$ and $\alpha = 0°$. This is because in this condition $k_x > k_y$. The infiltration of rainfall in the horizontal direction led to a dramatic rise of the underground water, which led to the decrease of the soil strength and the SF. What was contrary was that when $k_r = 100$ and $\alpha = 90°$, the SF reached the maximum (1.7), which was due to the low permeability coefficient in the horizontal direction (10^{-6} m/s), and the rainfall was difficult to infiltrate. It should be noticed that the SF increased when $k_r = 100$ and $\alpha = 15°$, which was because, in this situation, the underground water level decreased and the soil strength increased.

For clayey soil slope, rain water was difficult to infiltrate into the soil due to its low permeability, and the SF remained 1.7 under different k_r and α values, which was more stable than sandy soil slope under the same situations.

5. Conclusions

In this paper, the seepage characteristics and stability of slope in Luogang District, Guangzhou City, China were numerically simulated, considering the effect of sandy and clayey soil and the influence of the hydraulic conductivity anisotropy ratio k_r and direction α. The following conclusions can be obtained:

(1) The initial conditions are important for the subsequent calculation of unsaturated seepage. In this paper, the initial maximum suction of sand and clayey soil of −45 kPa was selected for numerical simulation, which was consistent with the actual situation.

(2) For sandy soil slope, the seepage characteristics and slope stability were greatly affected by the hydraulic conductivity anisotropy ratio k_r and direction α. The increase of k_r promoted the rainfall infiltration, which made the groundwater level and the surface water content rise. The decrease of α prevented rainfall infiltration, which made the rainfall hard to infiltrate. For clayey soil, the variations of k_r and α had little impact on the distribution of volumetric water content.

(3) The RID, RHG, and MWCS were defined to characterize the seepage response under rainfall. For sandy soil slope, the MWCS increased with the increase of k_r and α, and the RHG was −3–4 m. For clayey soil slope, the MWCS varied little and was higher than sandy soil slope, and the RID was 0.5–1 m.

(4) The minimum SF happened when $k_r = 100$ and $\alpha = 0°$, and varied dramatically with k_r and α for sandy soil slope. The SF for sandy soil slope was 1.3–1.7, while the SF for clayey soil slope was higher than that of sandy soil slope, and remained about 1.7.

(5) In actual engineering, it is necessary to consider the effect of the hydraulic conductivity anisotropy ratio k_r and direction α for sandy soil slope. However, for clayey soil slope, it can be treated as isotropic medium without considering its anisotropy for the purpose of simplifying the calculation.

Author Contributions: S.Y. completed the statistical analysis and wrote the paper; X.R., J.Z., and H.W. provided the writing ideas and supervised the study; Z.Z. did the numerical analysis. All authors have read and agreed to the published version of the manuscript.

Funding: This research was funded by The National Natural Science Fund (Grant No. U1765204).

Conflicts of Interest: The authors declare that there are no conflict of interest regarding the publication of this paper.

References

1.	Luo, G.C.; He, Z.M.; Luo, X.; Hu, Q.G. Numerical Simulation of Slope Excavation in Highway Renovation and Expansion Project. *Appl. Mech. Mater.* **2014**, *580*, 827–830. [CrossRef]
2.	Duong, T.T.; Do, D.M.; Yasuhara, K. Assessing the Effects of Rainfall Intensity and Hydraulic Conductivity on Riverbank Stability. *Water* **2019**, *11*, 741. [CrossRef]
3.	Liu, C.; Hao, W.; Wang, J.; Bai, W.; Qin, Z.; Li, Z. Effect of Connectivity Rate of Dominant Joints Extending Outward of the Slope on High Slope Stability. *J. Yangtze River Sci. Res. Inst.* **2014**, *31*, 74–77.
4.	Jotisankasa, A.; Mahannopkul, K.; Sawangsuriya, A. Slope Stability and Pore-Water Pressure Regime in Response to Rainfall: A Case Study of Granitic Fill Slope in Northern Thailand. *Geotech. Eng.* **2015**, *46*, 45–54.
5.	Yufei, K. Infinite Slope Stability under Transient Rainfall Infiltration Conditions. *Appl. Mech. Mater.* **2014**, *501*, 395–398.
6.	Oh, S.; Lu, N. Slope Stability Analysis under Unsaturated Conditions: Case Studies of Rainfall-induced Failure of Cut Slopes. *Eng. Geol.* **2014**, *184*, 96–103. [CrossRef]
7.	Zhang, Y.; Zhu, S.; Zhang, W.; Liu, H. Analysis of deformation characteristics and stability mechanisms of typical landslide mass based on the field monitoring in the Three Gorges Reservoir, China. *J. Earth Syst. Sci.* **2019**, *128*, 9. [CrossRef]
8.	Wu, L.; Wang, Z. Three Gorges Reservoir Water Level Fluctuation Influents on the Stability of the Slope's Analysis. *Adv. Mater. Res.* **2013**, *739*, 283–286. [CrossRef]
9.	Zhang, M.; Dong, Y.; Sun, P. Impact of reservoir impoundment-caused groundwater level changes on regional slope stability: A case study in the Loess Plateau of Western China. *Environ. Earth Sci.* **2012**, *66*, 1715–1725. [CrossRef]
10.	Chen, C.; Chen, H.; Wei, L.; Lin, G.; Iida, T.; Yamada, R. Evaluating the susceptibility of landslide landforms in Japan using slope stability analysis: A case study of the 2016 Kumamoto earthquake. *Landslides* **2017**, *14*, 1793–1801. [CrossRef]
11.	Wensong, W.; Guangzhi, Y.; Zuoan, W.; Qiangui, Z.; Guansen, C.; Xiaofei, J. Study of the dynamic stability of tailings dam based on time-history analysis method. *J. China Univ. Min. Technol.* **2018**, *47*, 271–279.
12.	Efremidis, G.; Avlonitis, M.; Konstantinidis, A.; Aifantis, E.C. A statistical study of precursor activity in earthquake-induced landslides. *Comput. Geotech.* **2017**, *81*, 137–142. [CrossRef]
13.	Duggan, A.R.; McCabe, B.A.; Goggins, J.; Clifford, E. An embodied carbon and embodied energy appraisal of a section of Irish motorway constructed in peatlands. *Constr. Build. Mater.* **2015**, *79*, 402–419. [CrossRef]
14.	Yu, S. Numerical Analysis of the Seepage Characteristics of Slopes with Weak Interlayers under Different Rainfall Levels. *Appl. Ecol. Environ. Res.* **2019**, *17*, 12465–12478. [CrossRef]
15.	Pande, R.K.; Uniyal, A. The fury of nature in Uttaranchal: Uttarkashi landslide of the year 2003. *Disaster Prev. Manag.* **2007**, *16*, 562–575. [CrossRef]
16.	Mo, P.-Q.; Yu, H.-S. Drained cavity expansion analysis with a unified state parameter model for clay and sand. *Can. Geotech. J.* **2017**, *55*, 1029–1040. [CrossRef]
17.	Fernando, P.; Yipei, W.; Xu, L. Acid rock drainage passive remediation using alkaline clay: Hydro-geochemical study and impacts of vegetation and sand on remediation. *Sci. Total Environ.* **2018**, *637*, 1262–1278.
18.	Amaral, L.F.; Vieira, C.M.F.; Delaqua, G.; Nicolite, M. Evaluation of Phyllite and Sand in the Heavy Clay Body Composition. *Mater. Sci. Forum* **2018**, *912*, 55–59. [CrossRef]
19.	Tan, Z.; Cai, M. Multi-factor sensitivity analysis of shallow unsaturated clay slope stability. *Int. J. Miner. Metall. Mater.* **2005**, *12*, 193–202.
20.	Qi, S.; Vanapalli, S.K. Influence of swelling behavior on the stability of an infinite unsaturated expansive soil slope. *Comput. Geotech.* **2016**, *76*, 154–169. [CrossRef]

21. Loretta, B.; Carastoian, A. Slope Stability Analysis Using the Unsaturated Stress Analysis. Case Study. *Procedia Eng.* **2016**, *143*, 284–291.

22. Frank, G.; Martin, F.; Albert, B. Effects of vegetation on the angle of internal friction of a moraine. *For. Snow Landsc. Res.* **2009**, *82*, 61–77.

23. Wu, X.Z. Probabilistic slope stability analysis by a copula-based sampling method. *Comput. Geosci.* **2013**, *17*, 739–755. [CrossRef]

24. Zhang, Y.; Chen, G.; Wang, B.; Li, L. An analytical method to evaluate the effect of a turning corner on 3D slope stability. *Comput. Geotech.* **2013**, *53*, 40–45. [CrossRef]

25. Shi, Z.; Shen, D.; Peng, M.; Zhang, L.; Wang, F.; Zhene, X. Slope stability analysis by considering rainfall infiltration in multi-layered unsaturated soils. *J. Hydraul. Eng.* **2016**, *47*, 977–985.

26. Weizao, W.; Qiang, X.; Guang, Z. Centrifugal model tests on sliding failure of gentle debris slope under rainfall. *Rock Soil Mech.* **2016**, *37*, 87–95.

27. Liu, J.; Zhang, J.; Feng, J. Green-Ampt Model for Layered Soils with Nonuniform Initial Water Content under Unsteady Infiltration. *Soil Sci. Soc. Am. J.* **2008**, *72*, 1041–1047. [CrossRef]

28. Mein, R.G.; Larson, C.L. Modeling infiltration during a steady rain. *Water Resour. Res* **1973**, *9*, 384–394. [CrossRef]

29. Chu, S.T. Infiltration during an unsteady rain. *Water Resour. Res.* **1978**, *14*, 461–466. [CrossRef]

30. Chen, L.; Young, M.H. Green-Ampt infiltration model for sloping surfaces. *Water Resour. Res.* **2006**, *42*, 887–896. [CrossRef]

31. Wu, C.P.; Luo, Y.S.; Chen, W. Indoor Model Experiment for Rainfall Effects on Bare Loess Slope Shape. *Bull. Soil Water Conserv.* **2013**, *01*, 121–125.

32. Li, L.; Luo, S.; Wang, Y.; Wei, W.; Li, C. Model tests for mechanical response of bedding rock slope UNDER different rainfall conditions. *Chin. J. Rock Mech. Eng.* **2014**, *33*, 755–762.

33. Zhang, Y. Study on Stability of Xiaokou Landslide under Different Rainfall Conditions. Ph.D. Thesis, SouthWest JiaoTong University, Chengdu, China, 2007.

34. Wang, H.; Chen, Z.X.; Zhang, D.M. Rock Slope Stability Analysis Based on FLAC3D Numerical Simulation. *Appl. Mech. Mater.* **2012**, *170*, 375–379.

35. Wang, X.; Li, Q.; Yu, Y. Simulation of the failure process of landslides based on extended finite element method. *Rock Soil Mech.* **2019**, *40*, 2435–2442.

36. Dou, H.; Han, T.; Gong, X. Reliability analysis of slope stability considering variability of soil saturated hydraulic conductivity under rainfall infiltration. *Rock Soil Mech.* **2016**, *37*, 1144–1152.

37. Song, Y.; Wu, Z.; Ye, G. Permeability and anisotropy of upper Shanghai clays. *Rock Soil Mech.* **2018**, *39*, 2139–2144.

38. Wang, T.; Yang, T.; Lu, J. Influence of dry density and freezing-thawing cycles on anisotropic permeability of loess. *Rock Soil Mech.* **2016**, *37*, 72–78.

39. Todd, D. *Groundwater Hydrology*; Jon Wiley & Sons Inc.: New York, NY, USA, 1980.

40. Yuan, J.-P.; Lin, Y.-L.; Ding, P.; Han, C.-L. Influence of anisotropy induced by fissures on rainfall infiltration of slopes. *Chin. J. Geotech. Eng.* **2016**, *38*, 76–82.

41. Zhao, Y.; Wang, W.; Huang, Y. Coupling analysis of seepage-damage-fracture in fractured rock mass and engineering application. *Chin. J. Geotech. Eng.* **2010**, *32*, 24–32.

42. Mahmood, K.; Ryu, J.; Kim, J.M. Effect of anisotropic conductivity on suction and reliability index of unsaturated slope exposed to uniform antecedent rainfall. *Landslides* **2013**, *10*, 15–22. [CrossRef]

43. Dong, J.J.; Tu, C.H.; Lee, W.R. Effects of hydraulic conductivity/strength anisotropy on the stability of stratified, poorly cemented rock slopes. *Comput. Geotech.* **2012**, *40*, 147–159. [CrossRef]

44. Yeh, H.F.; Tsai, Y.J. Analyzing the Effect of Soil Hydraulic Conductivity Anisotropy on Slope Stability Using a Coupled Hydromechanical Framework. *Water* **2018**, *10*, 905. [CrossRef]

45. Yizhao, W.; Yaohua, S. Influence of Rainfall Infiltration on Slope Stability at Shallow Layer. *J. Yangtze River Sci. Res. Inst.* **2017**, *34*, 122–125.

46. Zefeng, J.; Dayong, Z. Critical slip field of slope with tension crack under intensive rainfall. *Rock Soil Mech.* **2016**, *10*, 25–34.

47. Al-Juboori, M.; Datta, B. Performance evaluation of a genetic algorithm-based linked simulation-optimization model for optimal hydraulic seepage-related design of concrete gravity dams. *J. Appl. Water Eng. Res.* **2019**, *7*, 173–197. [CrossRef]

48. Al-Juboori, M.; Datta, B. Improved optimal design of concrete gravity dams founded on anisotropic soils utilizing simulation-optimization model and hybrid genetic algorithm. *ISH J. Hydraul. Eng.* **2019**, *36*, 1–18. [CrossRef]

49. Al-Juboori, M.; Datta, B. Reliability-based optimum design of hydraulic water retaining structure constructed on heterogeneous porous media: Utilizing stochastic ensemble surrogate model-based linked simulation optimization model. *Life Cycle Reliab. Saf. Eng.* **2019**, *8*, 65–84. [CrossRef]

50. GEO-SLOPE International Ltd. *Seepage Modeling with SEEP/W 2007*; Geo-Slope International Ltd.: Calgary, AB, Canada, 2010..

51. Fredlund, D.G.; Morgenstern, N.R.; Widger, R.A. The shear strength of unsaturated soils. *Can. Geotech. J.* **1978**, *15*, 313–321. [CrossRef]

52. Fredlund, D.G.; Rahardjo, H. *Soil Mechanics for Unsaturated Soils*; Wiley: New York, NY, USA, 1993.

53. Liu, J.; Zeng, L.; Fu, H. Variation law of rainfall infiltration depth and saturation zone of soil slope. *J. Cent. South Univ. (Sci. Technol.)* **2019**, *50*, 208–215.

54. Tang, D.; Li, D.; Zhou, C. Slope stability analysis considering antecedent rainfall process. *Rock Soil Mech.* **2013**, *34*, 3239–3248.

Article

Sensitivity Analysis on the Rising Relation between Short-Term Rainfall and Groundwater Table Adjacent to an Artificial Recharge Lake

Sheng-Hsin Hsieh [1,2], Li-Wei Liu [1], Wen-Guey Chung [1] and Yu-Min Wang [1,*]

[1] Department of Civil Engineering, National Pingtung University of Science and Technology, Pingtung 91201, Taiwan
[2] Department of Irrigation and Engineering, Council of Agriculture Executive Yuan, Taipei 10014, Taiwan
*   Correspondence: wangym@mail.npust.edu.tw; Tel.: +886-931-802-656

Received: 15 July 2019; Accepted: 14 August 2019; Published: 16 August 2019

Abstract: This study aimed to determine the highly sensitive variables for a groundwater simulation model adjacent to an artificial recharge lake (ARL) using short-term rainfall events. The model was established using an artificial neural network (ANN) with rainfall events. Normalized rainfall, rainfall intensity, and groundwater data were selected as model variables. The coefficient of determination (R^2) was used for model performance assessment. Finally, a sensitivity analysis (SA) was conducted to evaluate the importance of each model input. The study results indicated that the R^2 of the ANN model ranged between 0.759 and 0.914. The SA showed that the rainfall was more sensitive than rainfall intensity in the study area. Based on the SA results and relevant geological characteristics, it was observed that the rainfall of past 1-day, past 2-day, and past 3-day responded faster than the other variables to the wells near the river and the ARL. In addition, the past 2-day rainfall was highly sensitive to the groundwater table; this may be due to the fact that the well screen location was above sea level as observed in Wells 1, 2, and 6. The results indicate that the groundwater table variation is response-related to the distance from the wells to the river and the ARL, and the rainfall time-lag. This SA study is helpful to researchers wishing to study related ARL efficiency issues.

Keywords: groundwater recharge; infiltration; Lin-Bien River; rainfall intensity; artificial neural network; ANN

---

1. Introduction

In the 1970s, groundwater contributed 20% of the total water usage (TWU) in Taiwan, whereas faster economic growth with industrial expansion and population increase have greatly expanded water demand in Taiwan in the 1990s, the groundwater contributing 31% of the TWU in Taiwan [1,2]. In the 2010s, the groundwater water percentage of TWU has reached 34%, achieving 64.9% in Southern Taiwan [3,4]. However, due to anthropogenic interference and climate change, the water supply is extremely unstable, groundwater withdrawals may locally exceed recharge, and undesirable conditions may develop in the aquifer and in hydraulically connected surface waters [5]. Overexploited groundwater has led to seawater intrusion, land subsidence, lowering of groundwater levels, and salinization of soil, and has reduced the well water withdrawal yields in Taiwan [6,7]. Artificial recharge has been defined, as the process of replenishing groundwater through an artificial recharge lake (ARL), to increase groundwater storage in subterranean zones [8–10]. However, the process of groundwater recharge is complex. The infiltration impact is complicated by many confounding factors such as rainfall (or applied water) intensity, micro-topography, vegetation, soil texture, and vertical and horizontal heterogeneity in soil properties [11]. Morbidelli et al. indicated that, although a variety of local infiltration models for vertically homogeneous soils with constant initial soil water content and over horizontal surfaces have been proposed, the estimate of infiltration at different spatial scales (i.e., from the local to watershed scales) is a complex problem as further challenges are imposed by the natural spatial variability of soil hydraulic characteristics and that of rainfall [12]. Due to the intricate mechanism of groundwater, it is hard to simulate the hydrological phenomena in a hydro model because of the applied limitation, causing the use of extremely inconvenient models [13]. The simulation accuracy could be improved, by investing much time and many resources to establish a hydro model [14]. Thus, some researchers use statistical approaches to simulate groundwater variation. These approaches can reveal the stochastic dependence among the groundwater observations and their related variables, and commonly require a relatively fewer number of parameters than a physical-based model, with limited subjective assumptions [15]. Related research studies, such as those by Daliakopoulos et al. [16], Nayak et al. [17], Sahoo and Jha [18], and Liu et al. [19]. They used multiple linear regression (MLR) and artificial neural networks (ANNs) to establish the relationship using statistical models. For example, Daliakopoulos et al. used 17 years of data to establish ANN models according to time-lag rainfall, temperature, streamflow, and groundwater table in the Messara Valley basin (Greece). The dataset was divided into three parts for the purposes of training (11 years), cross-validation (3 years), and testing (3 years). The established ANNs included a feedforward neural network (FNN), a recurrent neural network (RNN), and a radial basis function network (RBFN). The coefficient of determination (R^2) was selected to evaluate model performance. The results showed that the R^2 in the FNN was between 0.592 and 0.993, in the RNN it was between 0.609 and 0.911, and in the RBFN it was 0.744 [16]. However, in statistical model development, it is more important to select highly related inputs to acquire accurate simulation results, and to exclude data having less impact on the output of the model, to reduce the freedom degree of the model and the processing. Many scholars use a sensitivity analysis (SA) as the determination method to obtain the relative importance of model inputs. Jha and Sahoo established multilayer perception (MLP), an RNN, and an RBFN to simulate groundwater variation in 17 sites in Jiangnan, Kochi, Japan. Groundwater table differences, past rainfall, stream stage, temperature, and seasonal dummy variables were selected as inputs. In that research, data were split into four years for the model training and validation, and the remaining two years of data for the model testing. The R^2 was selected for model performance evaluation. The results showed that, in 17 sites, the R^2 in MLP was between 0.781 and 0.971, in the RNN it was between 0.691 and 0.983, and in the RBFN it was between 0.691 and 0.983. After the ANN model was developed, an SA was conducted to identify the sensitivity of all inputs. The SA results showed that the H-4 well had the highest sensitivity for each input [20]. Ahlawat developed the relationship between precipitation and runoff using an ANN model in the Betwa catchment, India. A sensitivity analysis was conducted for model importance evaluation. The results indicated that some of the rainfall stations could be removed from the model because of the low sensitivity [21].

2. Materials and Methods

2.1. Study Area

The Lin-Bien catchment is located in Pingtung Plain, Taiwan (Figure 1, after [22]). There are three long-term record rainfall stations, namely, Taiwu-1, Xinlaiyi, and Nanhan, which were selected for rainfall data collection (Table 1). Groundwater monitoring wells included Wells 1 to 7, which are adjacent to the catchment boundary (Table 2). The location of each rainfall station and groundwater monitoring well is illustrated in Figure 2.

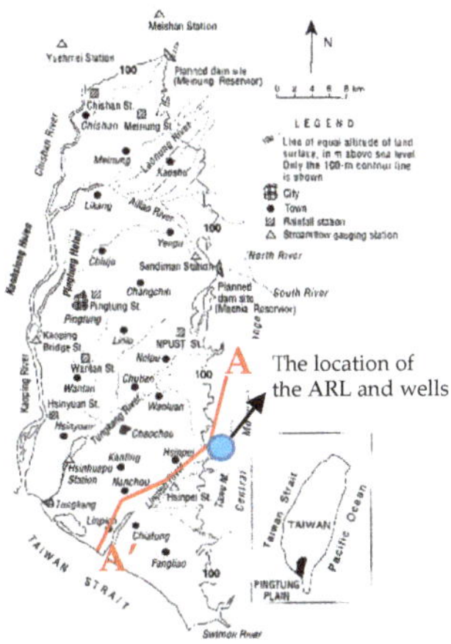

Figure 1. Location map of Pingtung Plain, after [22].

Table 1. Rainfall stations in the study catchment.

Rainfall Station (Period)	Annual Rainfall (mm)	Coordinate		Control Area	
		(TWD97-X)	(TWD97-Y)	(km^2)	(%)
Taiwu-1 (1955–2018)	4366.2	217,853.20	2,500,983.30	50.76	15
Xinlaiyi (1972–2018)	3646.9	216,084.71	2,492,152.60	141.13	40
Nanhan (1965–2018)	2506.0	211,948.60	2,481,928.20	156.97	45

Table 2. Information about each groundwater monitoring well in the study area.

Groundwater Monitoring Wells (Recording Period from 2010/05 to 2015/12)	Coordinate		Well Screen (Above Sea Level)
	(TWD97-X)	(TWD97-Y)	
Well 1	211,320.61	2,491,816.04	24 to 33
Well 2	211,084.75	2,490,990.15	14 to 23
Well 3	210,331.36	2,490,524.91	−17 to 5
Well 4	210,031.63	2,491,209.14	−18 to 6
Well 5	209,115.70	2,491,773.84	−17 to 5
Well 6	209,097.66	2,492,587.59	13 to 25
Well 7	210,184.92	2,492,093.04	−5 to 7

Ting et al. [23] point out that the groundwater catchment in the study area can be divided into three layers, that is, the upper layer, middle layer, and lower layer. The thickness of the upper layer, varying from 10 to 60 m, is assumed to be an unconfined aquifer. The aquifer of the fluvial deposits in the foothills consists of coarse sand and gravel, decreasing in size to sand in a downstream direction with a thickness in excess of 220 m. The middle layer, assumed to be semi-pervious, is a geological formation which has a very low transmissivity compared to the aquifer, with a thickness of about 20 m of clay and sandy clay. The lower layer is assumed to be unconfined in the upper reaches of the area and confined from the mid-fan; the storativity may thus alternate between confined and unconfined values. The aquifers are recharged directly by rainfall, river flow, and subsurface inflow from the northern upstream part of the fluvial fan. The study area is hydraulically bounded in the south by the sea (Taiwan Strait). The screens of each well were all set in the upper layer, which is mainly composed of gravel, and with slice sand, silt, and clay. The transmissivity, storativity, and maximum yield were determined as 9148 m^2/day, 6.5 × 10^{-3}, and 7084 m^3/day, respectively, by the Taiwan Provincial Groundwater Development Bureau [24]. The hydrogeological profile of each well in this study area is redrawn in Figure 3 [25], and the groundwater piezometric map of the Pingtung Plain's dry season is shown in Figure 4 [26].

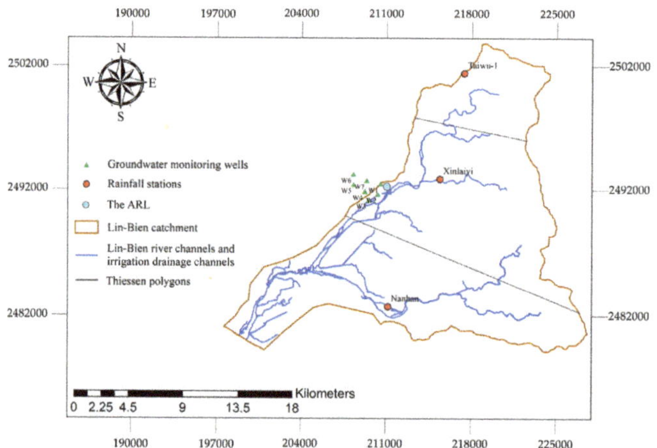

Figure 2. Location map of the study area (W1 to W7 is Well 1 to Well 7, respectively).

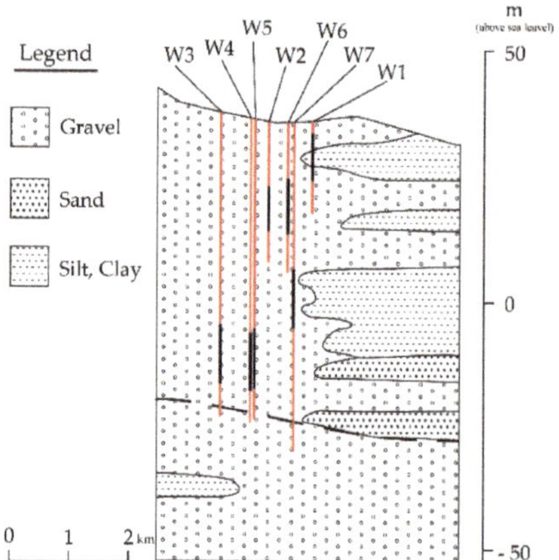

Figure 3. Hydrogeological profile of the study area (W1 to W7 is Well 1 to Well 7, respectively), after [25]. The A-A′ profile of Figure 1.

The infiltration rate in this study area was evaluated by infiltration experiment in a 30 × 30 × 1.63 m test pit in 2002 [27]. The initial infiltration rate was about 22.76 m/day and the average infiltration rate was 17.26 m/day. After a serial infiltration test, the difference in infiltration rate between 2003 and 2005 had a maximum variation from 15.2 to 10.33 m/day due to the sand addition experiment [28]. If one compares the ARL infiltration result to other studies, the infiltration rate is about 5.9 m/day in the Rokugo Alluvial Fan, Northern Japan [29]; Liu et al.'s study on underground reservoirs in the western suburbs of Beijing estimates the infiltration rate between 1.0 m/day and 3.6 m/day [30].

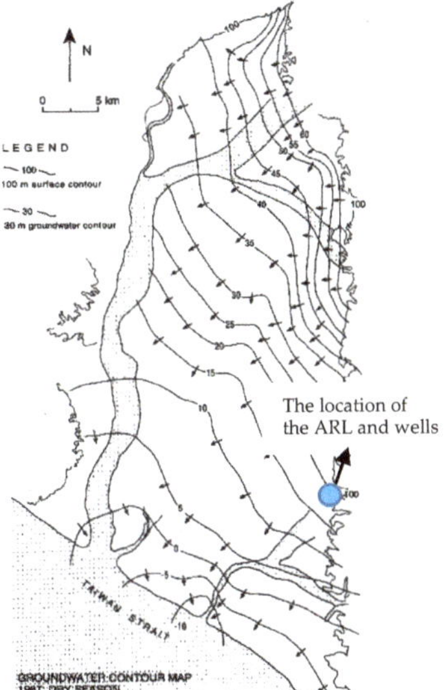

Figure 4. Groundwater piezometric map of the Pingtung Plain's dry season (November to May) [26].

2.2. Data Collection and Process

For this study, around five years of daily rainfall and daily groundwater table data collected from 2010/05 to 2015/12 were used. Before the model was established, original data underwent a pre-procedure process that included the following steps: (1) rainfall data processing, (2) groundwater table data processing, and (3) data normalization. The data collection and processing procedure steps are described in the following subsections.

2.2.1. Rainfall Data

Mean catchment rainfall can be calculated by several techniques, such as Thiessen polygon, isohyetal, polynomial, geostatistical, inverse distance weighted, multi-quadratic interpolation, and kriging [31,32]. This study referred to Tsou et al. [33] who used the Thiessen polygon method to calculate the mean catchment rainfall. The separated Thiessen's control area in the Taiwu-1 station was 50.76 km^2 (15%), in the Xinlaiyi station it was 141.13 km^2 (40%), and in the Nanhan station it was 156.97 km^2 (45%). Based on research of the effects of rainfall, rainfall intensity, and groundwater table variation [34], the groundwater table is more sensitive in a five-day time-lag in a shallow layer [35]. Hence, this study combined the rainfall of past 1-day (R1) to past 5-day (R5) and rainfall intensity of past 1-day (RI1) to past 5-day (RI5) as the model inputs.

2.2.2. Groundwater Data

This study referred to Nayak et al. using 2-day groundwater differences as output for the model establishment [17]. In addition, considering the hydrogeology in a shallow groundwater aquifer, the groundwater variation is significantly affected by 10 days of rainfall [36]. Therefore, this study separated rainfall and groundwater data by events, where groundwater table rising past 10 days until groundwater drop was regarded as an event.

2.2.3. Data Normalization

To prevent an error associated with extreme values, data normalization was conducted for the data regarding rainfall, rainfall intensity, and groundwater differences. In addition, this procedure randomized the data using the concept of stochastic statistics to understand the relative change amount in the database. Therefore, the data employed were bounded between 0 and 1 using Equation (1) below, and then they were reverted by following Equation (2) [37]:

$$x_{norm} = \frac{x - x_{min}}{x_{max} - x_{min}},$$

(1)

$$x = x_{norm} \times (x_{max} - x_{min}) + x_{min},$$

(2)

where x_{norm} is the normalized dimensionless variable, x is the observed value of the variable, x_{min} is the minimum value of the variable, and x_{max} is the maximum value of the variable.

2.3. Artificial Neural Network (ANN)

An artificial neural network is an algorithm for processing information by its dynamic state response to inputs. The neural network computes its output at each iteration (epoch) and compares it with the expected output of each input (exemplary) vector in order to calculate the error. An ANN comprises parallel systems that are composed of processing elements (PEs) or neurons, which are assembled in layers and connected through several links or weights. After feeding input data to the input layer, they pass through and are operated on by the network until the output is produced at the output layer. Each neuron receives numerous inputs from other neurons through some weighted connections. These weighted inputs are then summed, and a standard threshold is added, generating the argument for a transfer function (usually linear, logistic, or hyperbolic tangent), which in turn produces the final output of the neuron [38,39].

Hsieh et al. and Liao et al. compared MLP, a time lag recurrent network (TLRN), and a time delay neural network (TDNN) for groundwater simulation in the Lin-Bien River catchment. The MLP model is appropriate and suitable for this study area [40,41]. In order to learn more complex decision functions, inputs are fed into a number of perceptron nodes, each with its own set of weights and threshold [42]. The outputs of these nodes are then inputted into another layer of nodes, and so on. The output of the final layer of nodes is the output of the network. Such a network is termed MLP, and the layers of nodes whose input and output are seen only by other nodes are termed hidden [43]. The connection weights are computed by means of a learning algorithm. There are different variants of back-propagation learning algorithms in the literature [44]. The illustration of the MLP model is shown in Figure 5.

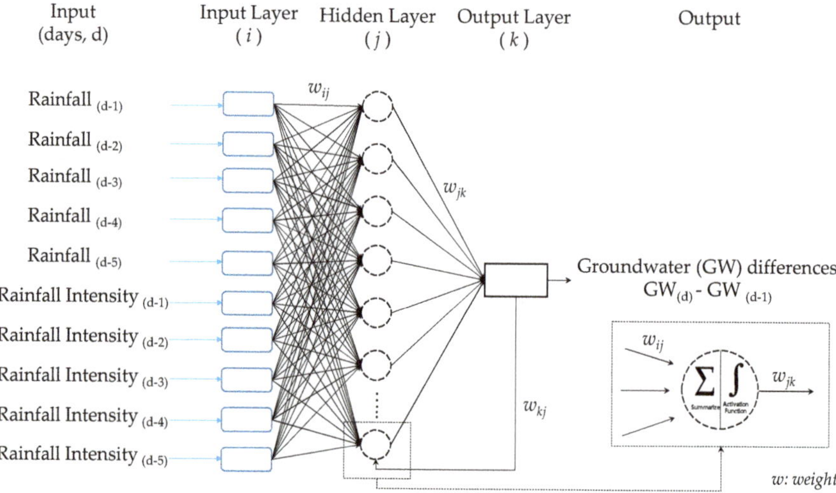

Figure 5. Illustration of the multilayer perception (MLP) model, after [39].

2.4. Sensitivity Analysis (SA)

A sensitivity analysis is a tool for model input importance assessment. Referring to Jha and Sahoo [20] and Memarian et al. [45,46], the SA calculation equation (Equation (3)) is given as follows [47]. When the change is small, the input is less sensitive in the model; on the contrary, when the changes are significant, the input is highly sensitive:

$$S_k = \frac{\sum\limits_{p=1}^{p} \sum\limits_{n=1}^{n} \left(y_{ip} - \overline{y_{ip}} \right)}{a_k^2} \tag{3}$$

where S_k is the sensitivity index for input k, $\overline{y_{ip}}$ is the ith output obtained with the fixed weights for the pth pattern, n is the number of network outputs, p is the number of patterns, and α_k^2 is the variance of the input k.

3. Results

The ANN model was established using normalized rainfall and groundwater table data. The R^2 was selected for model performance evaluation. Thus, the sensitivity analysis was conducted to evaluate the importance of the input for the model simulation. All results are discussed below.

3.1. ANN Results

Referring to the training, validation, and testing arrangements from Tayfur and Singh [48], Yang et al. [49] and Memarian et al. [45], respectively, this study used 70% of the data for training, 20% for validation, and 10% for testing. With respect to the PE and hidden layer selection, Cheung et al. suggest 10 PEs in 1 hidden layer [50]. Thus, in this study, PEs were set at 10, and the hidden layer was set at 1. The simulated results revealed that the R^2 of the MLP model in Well 1 was 0.848, in Well 2 it was 0.854, in Well 3 it was 0.914, in Well 4 it was 0.897, in Well 5 it was 0.759, in Well 6 it was 0.841, and in Well 7 it was 0.812. All MLP model values showed high correlation ($R^2 > 0.8$) except for Well 5, which was the farthest well from the Lin-Bien River and the ARL. Comparing Well 5 with a similar location well (Well 6), the well screen of Well 6 (13–25 m, above sea level) was shallower than that of Well 5 (−17 to 5 m, above sea level). The MLP model performance values (R^2) are shown in Table 3. The model testing correlation relationship of each well is presented in Figures 6–12.

Table 3. MLP model performance.

Groundwater Monitoring Wells	R^2
Well 1	0.848
Well 2	0.854
Well 3	0.914
Well 4	0.897
Well 5	0.759
Well 6	0.841
Well 7	0.812

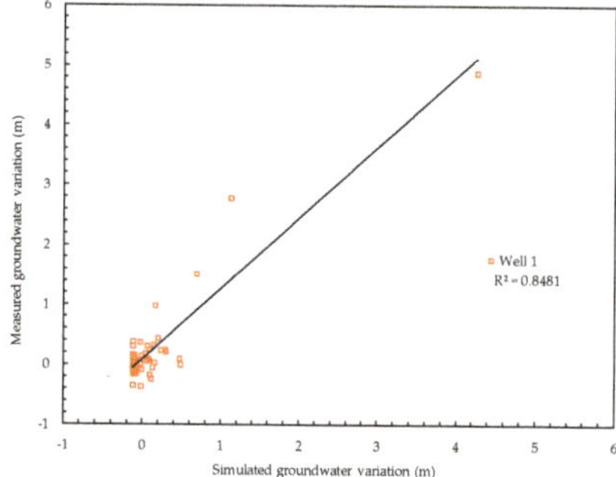

Figure 6. The model testing correlation relationship of Well 1.

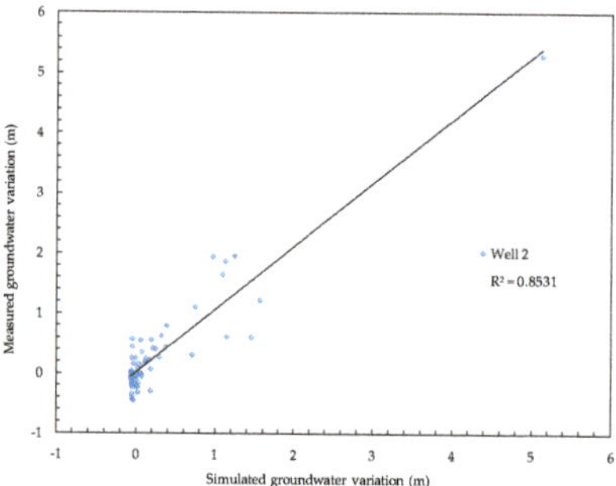

Figure 7. The model testing correlation relationship of Well 2.

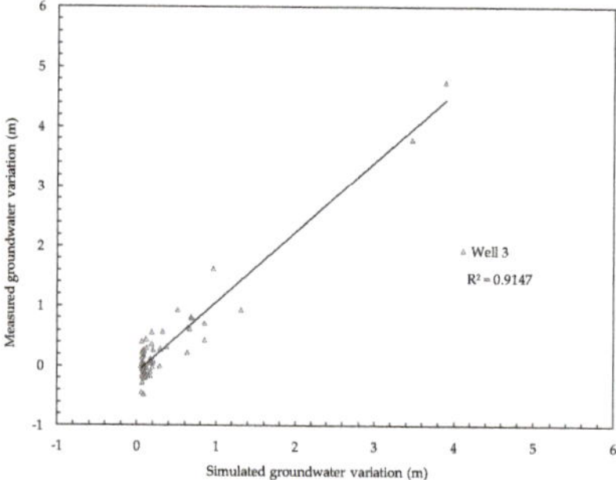

Figure 8. The model testing correlation relationship of Well 3.

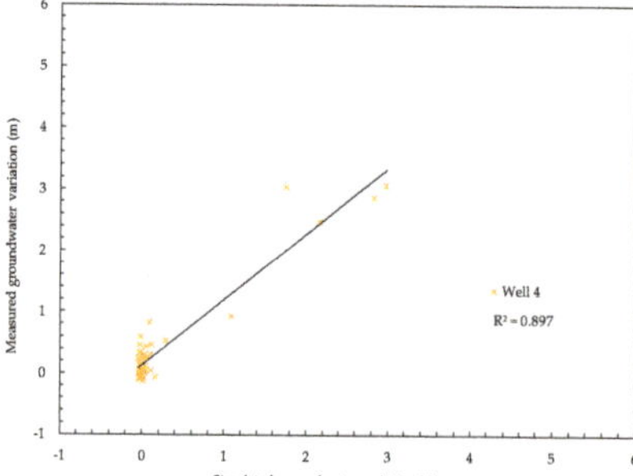

Figure 9. The model testing correlation relationship of Well 4.

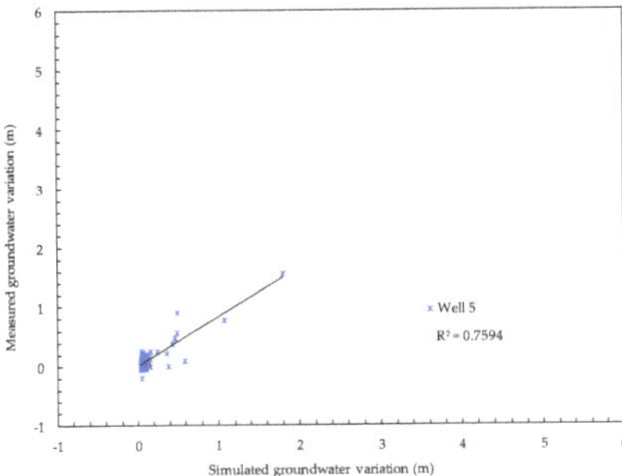

Figure 10. The model testing correlation relationship of Well 5.

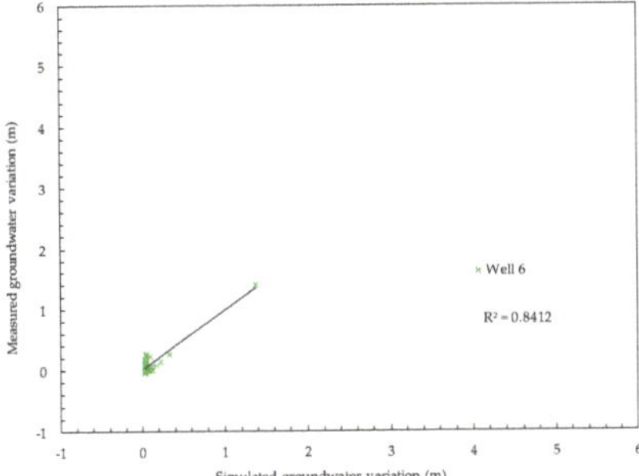

Figure 11. The model testing correlation relationship of Well 6.

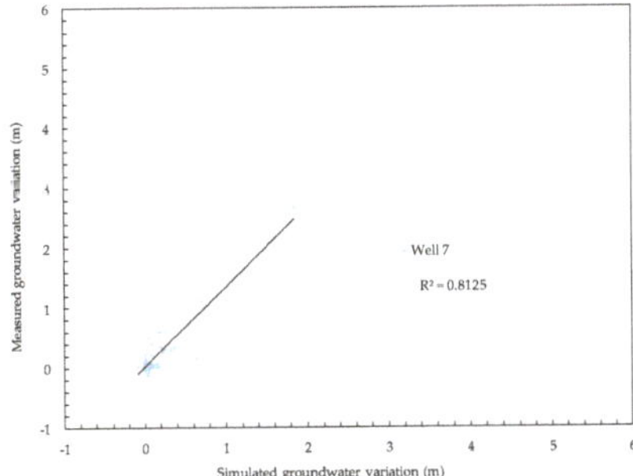

Figure 12. The model testing correlation relationship of Well 7.

3.2. SA Results

A sensitivity analysis was performed on the MLP model's simulation results to understand the relative sensitivity of each input to the output. All SA results are shown in Table 4 and Figure 13. Referring to Jha and Sahoo, the level of sensitivity was classified into five categories [20], as shown in Table 5. Based on the value of the sensitivity index for a particular input at the well, the influence level of the model's sensitivity to the input was ranked on a scale of 1 to 5 (Table 6). The sensitivity analysis indicated that, for past 5-day rainfall and rainfall intensity, R2 and RI3 were the highest and second highest sensitivities in Well 1; in Well 2, the results were R2 and R3; in Well 3, they were R4 and R2; in Well 4, they were RI2 and RI1; in Well 5, they were R5 and R4; in Well 6, they were RI4 and R2; and in Well 7, they were R4 and R5. The rainfall (R) is more sensitive than rainfall intensity (RI) in this research area. The highest and second highest sensitivities are listed in Table 7 (if the ranked result in the same value, the original SA index was compared).

Table 4. Sensitivity analysis index.

Inputs	Well 1	Well 2	Well 3	Well 4	Well 5	Well 6	Well 7
RI5	0.0216	0.0207	0.0163	0.0064	0.0096	0.0049	0.0145
RI4	0.0045	0.0025	0.0035	0.0028	0.0047	0.0133	0.0033
RI3	0.0223	0.0220	0.0156	0.0101	0.0076	0.0002	0.0164
RI2	0.0040	0.0157	0.0231	0.0650	0.0044	0.0102	0.0037
RI1	0.0017	0.0163	0.0130	0.0212	0.0051	0.0021	0.0200
R5	0.0053	0.0132	0.0181	0.0035	0.0238	0.0084	0.0498
R4	0.0203	0.0324	0.0639	0.0029	0.0181	0.0069	0.1489
R3	0.0136	0.0594	0.0343	0.0036	0.0048	0.0036	0.0443
R2	0.0849	0.1095	0.0362	0.0108	0.0030	0.0127	0.0092
R1	0.0102	0.0027	0.0048	0.0173	0.0060	0.0042	0.0372

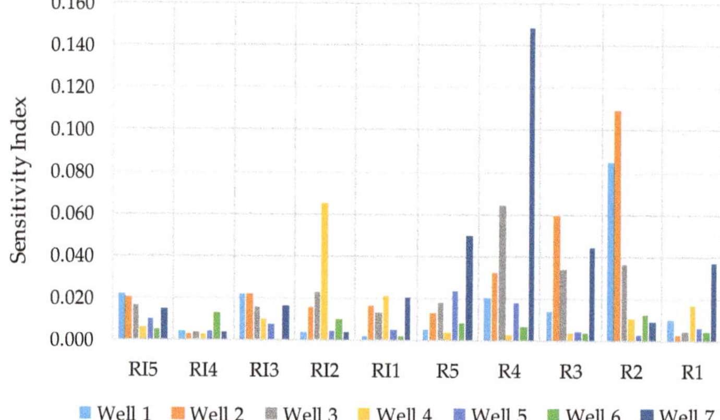

Figure 13. Sensitivity index of each input of the MLP model.

Table 5. Categorization of sensitivity and associated ranks.

Category of Sensitivity	Value of Sensitivity Index	Rank
1. Very high sensitivity	>0.1	1
2. High sensitivity	0.05 to 0.1	2
3. Moderate sensitivity	0.01 to 0.05	3
4. Low sensitivity	0.005 to 0.01	4
5. Very low sensitivity	≤0.005	5

Table 6. Ranked sensitivity analysis index of each input.

Inputs	Well 1	Well 2	Well 3	Well 4	Well 5	Well 6	Well 7
RI5	3	3	3	4	4	5	3
RI4	5	5	5	5	5	3	5
RI3	3	3	3	3	4	5	3
RI2	5	3	3	2	5	3	5
RI1	5	3	3	3	4	5	3
R5	4	3	3	5	3	4	3
R4	3	3	2	5	3	4	1
R3	3	2	3	5	5	5	3
R2	2	1	3	3	5	3	4
R1	3	5	5	3	4	5	3

Table 7. Highest and 2nd highest sensitivities.

Groundwater Monitoring Wells	Highest Sensitivity		2nd Highest Sensitivity	
	Index	Variable	Index	Variable
Well 1	2	R2	3	RI3
Well 2	1	R2	2	R3
Well 3	2	R4	3	R2
Well 4	2	RI2	3	RI1
Well 5	3	R5	3	R4
Well 6	3	RI4	3	R2
Well 7	1	R4	3	R5

When considering the SA results and the information in the location map of the study area (Figure 2), one can see that Wells 1 to 3 are within the study catchment, which is adjacent to Lin-Bien River and the ARL, and responded faster in R1, R2, R3, RI1, RI2, RI3, respectively; Wells 4 to 7 are outside the catchment and far from the ARL; R4, R5, RI4, RI5 were found to be sensitive in Wells 5 to 7. It can be concluded that the groundwater table variation adjacent to the Lin-Bien River and the ARL is response-related with rainfall time-lag. It is worth noting that, although Well 4 is located outside the catchment, it still has a certain degree of sensitivity from past 3-day rainfall (R1, R2, R3, RI1, RI2, RI3, respectively), the reason for which is discussed below.

The sensitivity of each well is influenced by the well screen position. The reason for past 3-day rainfall (R1, R2, R3, RI1, RI2, RI3, respectively) being sensitive in Well 4 is that the well screen's location is similar to the one in Well 3. After considering the SA results about the information for each groundwater monitoring well (Table 2), it was observed that 2-day rainfall (R1, R2, RI1, RI2, respectively) was more sensitive due to the well screen location being above sea level, for example, in Wells 1, 2, and 6.

4. Discussion

The purpose of this study was to discover the highly sensitive variables of a groundwater simulation model adjacent to an ARL using short-term rainfall. The model was established by MLP with past 10-day short-term events. Normalized rainfall (R1 to R5), rainfall intensity (RI1 to RI5), and groundwater data were selected as model variables. The R^2 was used for model accuracy assessment [18]. A sensitivity analysis (SA) was conducted to evaluate the importance of each model input. The study results are discussed as follows.

First, during the model establishment, the MLP model was suggested for development [5,40,41]. The best fit R^2 of the MLP model was 0.914 (Well 3) and the lowest R^2 in MLP was 0.759 (Well 5). This model uses rainfall as an input; the rainfall converges in the ARL and the river then infiltrates into the groundwater layer. The farthest well (Well 5) from the Lin-Bien River and the ARL responded more slowly than the other wells with a lower R^2. Comparing Well 5 with a similar location well (Well 6), the well screen of Well 6 (13–25 m, above sea level) was shallower than that of Well 5 (−17 to 5 m, above sea level). Therefore, the rainfall response in Well 6 was faster than in Well 5.

Second, a sensitivity analysis was conducted for input importance evaluation. The highest and second highest inputs in Well 1 were R2 and RI3; in Well 2, they were R2 and R3; in Well 3, they were R4 and R2; in Well 4, they were RI2 and RI1; in Well 5, they were R5 and R4; in Well 6, they were RI4 and R2; and in Well 7, they were R4 and R5. The rainfall amount (R) was more sensitive than rainfall intensity (RI) in this research area.

Finally, when considering the SA results for the Lin-Bien River catchment area (Figure 2) and the relevant hydrogeological profile (Figure 3), a slight difference between "within the catchment" and "outside the catchment" can be found. Wells 1 to 3 are in the study catchment and adjacent to Lin-Bien River and the ARL, and responded faster in past 3-day rainfall (R1, R2, R3, RI1, RI2, RI3, respectively) than the other four wells. On the other hand, R4, R5, RI4, RI5 were found to be sensitive in Wells 5 to 7.

Although Well 4 was located outside the catchment, it still had a certain degree of sensitivity from past 3-day rainfall (R1, R2, R3, RI1, RI2, RI3, respectively) because the location of its well screen was similar to the one in Well 3. Moreover, R2 was highly sensitive due to the well screen being located above sea level, that is, Wells 1, 2, and 6. It can be concluded that the groundwater table variation in this study area is response-related with the distance from the wells to the river and the ARL, and the rainfall time-lag.

5. Conclusions

A sensitivity analysis is not only a tool for the input importance evaluation of a groundwater simulation model, but it is also a useful method, considered with a hydrogeological map and geological characteristics, for explaining influences on the physical mechanism of groundwater infiltration, taking into account the complexity of the relevant physical groundwater model's development. This SA study is helpful to researchers wishing to study related ARL efficiency issues.

Author Contributions: The authors contributed equally in preparing this manuscript. They are all well conversant with its content and have agreed to the sequence of the authorship. S.-H.H. initiated and provided the concept and monitored the groundwater table variation. L.-W.L. conducted the data analysis. W.-G.C. reviewed and edited the manuscript. Y.-M.W. supervised the analyzing work, and provided oversight for the analysis of data and editing of the manuscript.

Funding: This research was funded by the Pingtung County Government, Taiwan, under the grant of "2015 Hydrological Data Analysis and Evaluation of the Great Chaozhou Artificial Groundwater Recharge Lake Implementation Plan" and partially funded by MOST, Taiwan, grant number 105-2221-E-020-009.

Acknowledgments: The authors gratefully acknowledge the team of Hydraulic Laboratory of NPUST for the field observation and data analysis. Also to acknowledge the contact person Wei-Tai Lee for his sincere help.

Conflicts of Interest: The authors declare no conflict of interest.

References

1. Peng, T.R.; Lu, W.C.; Chen, K.Y.; Zhan, W.J.; Liu, T.K. Groundwater-recharge connectivity between a hills-and-plains' area of western Taiwan using water isotopes and electrical conductivity. *J. Hydrol.* **2014**, *517*, 226–235. [CrossRef]

2. Wang, C.H.; Kuo, C.H.; Chang, F.C. The Changing Face of the Groundwater Environment in Taiwan. *Bull. Cent. Geol. Surv.* **2004**, *17*, 1–22.

3. T.W.R.A. *Taiwan Water Usage Annual Report*; Taiwan Water Resources Agency, Ministry of Economic Affairs: Taipei, Taiwan, 2017; p. A-14.

4. T.W.R.A. Taiwan Groundwater Withdrawal and Pumping. Available online: https://www.wra.gov.tw/6950/7170/7356/7488/13319 (accessed on 14 July 2019).

5. Liao, F.L. Using Artificial Neural Network for Estimating the Effects on Groundwater from Artificial Recharge or Rainfall. National Pingtung University of Science and Technology: Pingtung County, Taiwan, 2017.

6. T.P.W.C.B. *Compilation and Analysis of Existing Groundwater and Subsidence Data of Taiwan II–the Pingtung Plain*; Ministry of Economic Affairs: Taipei, Taiwan, 1994.

7. T.P.W.C.B. *Study on the Improvement of the Groundwater Monitoring System in the Pingtung Plain*; Ministry of Economic Affairs: Taipei, Taiwan, 1998.

8. Ineson, J. Hydrogeological and Groundwater Aspects of Artificial Recharge. In Proceedings of the Artificial Groundwater Recharge Conference, England, UK, 21–24 September 1970; pp. 1–14.

9. Muckel, D. Artificial Recharge in Relation to Groundwater Storage. In Proceedings of the Annual Conference on Water for Texas, College Station, TX, USA, 16–18 September 1958; pp. 85–94.

10. Walton, W.C. Groundwater Resource Evaluation. In *McGraw-Hill Series in Water Resources and Environmental Engineering (USA) eng*; McGraw-Hill: New York, NY, USA, 1970.

11. Morbidelli, R.; Saltalippi, C.; Flammini, A.; Cifrodelli, M.; Picciafuoco, T.; Corradini, C.; Govindaraju, R. Laboratory investigation on the role of slope on infiltration over grassy soils. *J. Hydrol.* **2016**, *543*, 542–547. [CrossRef]

12. Morbidelli, R.; Corradini, C.; Saltalippi, C.; Flammini, A.; Dari, J.; Govindaraju, R. A New Conceptual Model for Slope-Infiltration. *Water* **2019**, *11*, 678. [CrossRef]

13. Chen, C.T.; Chen, L.S.; Huang, J.Y.; Huang, S.J. A Study on Ground Water Level Forecasting by Combining Neural Networks and Semi-variogram Model. *J. Taiwan Agric. Eng.* **2015**, *61*, 14–28. [CrossRef]

14. Coppola, E.A., Jr.; Rana, A.J.; Poulton, M.M.; Szidarovszky, F.; Uhl, V.W. A neural network model for predicting aquifer water level elevations. *Groundwater* **2005**, *43*, 231–241. [CrossRef] [PubMed]

15. Yu, H.-L.; Lin, Y.-C. Analysis of space–time non-stationary patterns of rainfall–groundwater interactions by integrating empirical orthogonal function and cross wavelet transform methods. *J.Hydrol.* **2015**, *525*, 585–597. [CrossRef]

16. Daliakopoulos, I.N.; Coulibaly, P.; Tsanis, I.K. Groundwater level forecasting using artificial neural networks. *J. Hydrol.* **2005**, *309*, 229–240. [CrossRef]

17. Nayak, P.C.; Rao, Y.S.; Sudheer, K. Groundwater level forecasting in a shallow aquifer using artificial neural network approach. *Water Resour. Manag.* **2006**, *20*, 77–90. [CrossRef]

18. Sahoo, S.; Jha, M.K. Groundwater-level prediction using multiple linear regression and artificial neural network techniques: A comparative assessment. *Hydrogeol. J.* **2013**, *21*, 1865–1887. [CrossRef]

19. Liu, L.W.; Hsieh, S.H.; Chung, W.G.; Wang, Y.M. Sensitivity Analysis on the Rising Relation of Short-term Rainfall and Unconfined Aquifer Groundwater Table. In Proceedings of the 2017 4th International Conference on Coastal and Ocean Engineering (ICCOE 2017), Osaka, Japan, 29 March 2017; p. 64.

20. Jha, M.K.; Sahoo, S. Efficacy of neural network and genetic algorithm techniques in simulating spatio-temporal fluctuations of groundwater. *Hydrol. Process.* **2015**, *29*, 671–691. [CrossRef]

21. Ahlawat, R. Hydrological Data Network Modelling Using Artificial Neural Network in Betwa Catchment. *Int. J. Soft Comput. Eng. (IJSCE)* **2014**, *3*, 132–134.

22. Ting, C.S.; Kerh, T.; Liao, C.J. Estimation of groundwater recharge using the chloride mass-balance method, Pingtung Plain, Taiwan. *Hydrogeol. J.* **1998**, *6*, 282–292. [CrossRef]

23. Ting, C.S.; Zhou, Y.; Vries, J.D.; Simmers, I. Development of a Preliminary Ground Water Flow Model for Water Resources Management in the Pingtung Plain, Taiwan. *Groundwater* **1998**, *36*, 20–36. [CrossRef]

24. T.P.W.C.B. *Investigation Report on Ground Water Resources of the Pingtung Plain*; Ministry of Economic Affairs: Taichung, Taiwan, 1961.

25. Ministry of Economic Affairs. *Geological Sensitive Groundwater Recharge Area Delineation Plan-G0002 Pingtung Plain*; Ministry of Economic Affairs: Taipei, Taiwan, 2014.

26. Ting, C.S. Groundwater Resources Evaluation and Management for Pingtung Plain, Taiwan. Ph.D. Thesis, Vrije Universiteit, Amsterdam, The Netherlands, 1997.

27. C.T.C.I. *Pingtung County Water Resources Development and Conservation of The Overall Planning and Lin-Bien River Upstream Artificial Lake Engineering Project Settings–Groundwater Recharge Field Test Report*; Pingtung County Government: Pingtung, Taiwan, 2000; pp. 3–23.

28. Tu, Y.-C.; Ting, C.-S.; Tsai, H.-T.; Chen, J.-W.; Lee, C.-H. Dynamic analysis of the infiltration rate of artificial recharge of groundwater: A case study of Wanglong Lake, Pingtung, Taiwan. *Environ. Earth Sci.* **2011**, *63*, 77–85. [CrossRef]

29. Hida, N.; Ishikawa, E.; Ohta, Y. Experimental Study of Basin Artificial Recharge of Ground Water in Rokugo Alluvial Fan, Northern Japan. *J. Groundw. Hydrol.* **1999**, *41*, 23–33. [CrossRef]

30. Liu, J.X.; Cai, Q.S. *Research on the Groundwater Reservoir in the West Suburb of Beijing*; Geological Publishing Company: Beijing, China, 1988.

31. Abo-Monasar, A.; Al-Zahrani, M.A. Estimation of rainfall distribution for the southwestern region of Saudi Arabia. *Hydrol. Sci. J.* **2014**, *59*, 420–431. [CrossRef]

32. Ewea, H.A.; Elfeki, A.M.M.; Bahrawi, J.A.; Al-Amri, N.S. Sensitivity analysis of runoff hydrographs due to temporal rainfall patterns in Makkah Al-Mukkramah region, Saudi Arabia. *Arab. J. Geosci.* **2016**, *9*, 424. [CrossRef]

33. Tsou, I.; Chen, C.N.; Wang, Y.M.; Kuo, Y.C. A Simulation Study for Flood Disaster Reduction by Detention Pond in Lin-Bian River Basin. In Proceedings of the International Conference on Disaster Management, Kumamoto, Japan, 25 August 2012; pp. 399–406.

34. Bogaard, T.A. Analysis of Hydrological Processes in Unstable Clayey Slopes. Ph.D. Thesis, Utrecht University, Utrecht, The Netherlands, 2001.

35. Jasmin, I.; Murali, T.; Mallikarjuna, P. Statistical analysis of groundwater table depths in upper Swarnamukhi River basin. *J. Water Resour. Prot.* **2010**, *2*, 577. [CrossRef]

36. Apps, J.A.; Zheng, L.; Spycher, N.; Birkholzer, J.T.; Kharaka, Y.; Thordsen, J.; Kakouros, E.; Trautz, R. Transient changes in shallow groundwater chemistry during the MSU ZERT CO_2 injection experiment. *Energy Procedia* **2011**, *4*, 3231–3238. [CrossRef]

37. Wang, Y.-M.; Elhag, T.M. An adaptive neuro-fuzzy inference system for bridge risk assessment. *Expert Syst. Appl.* **2008**, *34*, 3099–3106. [CrossRef]

38. Talebizadeh, M.; Morid, S.; Ayyoubzadeh, S.A.; Ghasemzadeh, M. Uncertainty analysis in sediment load modeling using ANN and SWAT model. *Water Resour. Manag.* **2010**, *24*, 1747–1761. [CrossRef]

39. Traore, S.; Wang, Y.M.; Chung, W.G. Predictive accuracy of backpropagation neural network methodology in evapotranspiration forecasting in Dédougou region, western Burkina Faso. *J. Earth Syst. Sci.* **2014**, *123*, 307–318. [CrossRef]

40. Hsieh, S.H.; Chung, W.G.; Wang, Y.M. Using Artificial Neural Networks for Groundwater Table in Unconfined Aquifer Forecasting. In Proceedings of the 2017 4th International Conference on Coastal and Ocean Engineering (ICCOE 2017), Osaka, Japan, 29 March 2017; p. 63.

41. Liao, F.L.; Liu, L.W.; Chung, W.G.; Wang, Y.M. Using ANN for modeling the unconfined groundwater table variation induced by artificial recharge lake in dry season. In Proceedings of the 2017 4th International Conference on Coastal and Ocean Engineering (ICCOE 2017), Osaka, Japan, 29 March 2017; p. 98.

42. Bishop, C.M. *Neural Networks for Pattern Recognition*; Oxford University Press: New York, NY, USA, 1995.

43. Lippmann, R.P. An introduction to computing with neural nets. *IEEE Assp Mag.* **1987**, *4*, 4–22. [CrossRef]

44. Hagan, M.T.; Demuth, H.B.; Beale, M.H.; De Jesús, O. *Neural Network Design*; Pws Pub.: Boston, MA, USA, 1996; Volume 20.

45. Memarian, H.; Balasundram, S.K.; Tajbakhsh, M. An expert integrative approach for sediment load simulation in a tropical watershed. *J. Integr. Environ. Sci.* **2013**, *10*, 161–178. [CrossRef]

46. Memarian, H.; Bilondi, M.P.; Rezaei, M. Drought prediction using co-active neuro-fuzzy inference system, validation, and uncertainty analysis (case study: Birjand, Iran). *Theor. Appl. Climatol.* **2016**, *125*, 541–554. [CrossRef]

47. Principe, J.C.; Euliano, N.R.; Lefebvre, W.C. *Neural and Adaptive Systems: Fundamentals through Simulations*; Wiley: New York, NY, USA, 2000; Volume 672.

48. Tayfur, G.; Singh, V.P. ANN and fuzzy logic models for simulating event-based rainfall-runoff. *J. Hydraul. Eng.* **2006**, *132*, 1321–1330. [CrossRef]

49. Yang, C.T.; Marsooli, R.; Aalami, M.T. Evaluation of total load sediment transport formulas using ANN. *Int. J. Sediment Res.* **2009**, *24*, 274–286. [CrossRef]

50. Cheung, S.O.; Wong, P.S.P.; Fung, A.S.; Coffey, W. Predicting project performance through neural networks. *Int. J. Proj. Manag.* **2006**, *24*, 207–215. [CrossRef]

Article

A Sensitivity Analysis of Simulated Infiltration Rates to Uncertain Discretization in the Moisture Content Domain

Lulu Liu [1] and Han Yu [2,3,*]

[1] Department of Mathematics, School of Science, Nanjing University of Science and Technology, Nanjing 210094, China; lulu.liu@njust.edu.cn

[2] School of Computer Science and Technology, Nanjing University of Posts and Telecommunications, Nanjing 210023, China

[3] Division of Physical Science and Engineering, King Abdullah University of Science and Technology, Thuwal 23955-6900, Saudi Arabia

* Correspondence: han.yu@njupt.edu.cn

Received: 25 April 2019; Accepted: 4 June 2019; Published: 7 June 2019

Abstract: An unconditionally mass conservative hydrologic model proposed by Talbot and Ogden provides an effective and fast technique for estimating region-scale water infiltration. It discretizes soil moisture content into a proper but uncertain number of hydraulically interacting bins such that each bin represents a collection of pore sizes. To simulate rainfall-infiltration, a two-step alternating process runs until completion; and these two steps are surface water infiltration into bins and redistribution of inter-bin flow. Therefore, a nonlinear dynamical system in time is generated based on different bin front depths. In this study, using rigorous mathematical analysis first reveals that more bins can produce larger infiltration fluxes, and the overall flux variation is nonlinear with respect to the number of bins. It significantly implies that a greater variety of pore sizes produces a larger infiltration rate. An asymptotic analysis shows a finite change in infiltration rates for an infinite number of bins, which maximizes the heterogeneity of pore sizes. A corollary proves that the difference in the predicted infiltration rates using this model can be quantitatively bounded under a specific depth ratio of the deepest to the shallowest bin fronts. The theoretical results are demonstrated using numerical experiments in coarse and fine textured soils. Further studies will extend the analysis to the general selection of a suitable number of bins.

Keywords: infiltration; moisture content discretization; uncertain; bin

---

1. Introduction

The recharge of groundwater [1,2] is critical in many aspects, for example, natural environments, industry, and agriculture. Therefore, recharging aquifers is urgent in regions with growing demands [3,4] on water supplies that are the key to the local ecosystem and to economic development [5,6]. By using either natural or artificial methods to conduct the recharge, estimating infiltration is usually subject to both uncertainties and multiple types of errors. Moreover, the problem of rainfall-induced shallow landslides represents the most common natural hazard [7] in some areas of the world. These landslides are activated by intense rainfall events where water infiltration causes an increase of both volumetric content and pore pressure, thus worsening the slope stability in landslide prone areas. Thus, effective and economic numerical models are first needed to simulate the movement of water in the vadose zone, especially for large-scale distributed hydrologic applications over a relatively long period [8,9].

For groundwater infiltration there are advanced reservoir modeling methods based on the Richards equation [10] that can provide exact solutions to estimate infiltration [11,12]. However, these methods are computationally expensive especially for a large geographic area such as a city or a large ranch or farm. To simulate the rain-infiltration process over a number of years, they are complicated and computationally costly since they determine where the water moves in space. The Green–Ampt model [13], due to its computational convenience, is widely used in estimating infiltration parameters and states, such as flux, accumulative water content, and infiltration time [14–17]. Nevertheless, it makes some ideal, although unrealistic assumptions, for instance, it assumes the existence of an abrupt wetting front, and uniform water content behind the wetting front. To avoid these limitations, several contributions have modified this model. They can be classified as experimentally based corrections [18,19], and mathematically or physically based optimizations [20–22]. To avoid the drawbacks of the Green–Ampt model and rapidly determine how well ground water and aquifers are recharged only, the Talbot–Ogden model is proposed [23,24] for estimating large-scale surface water infiltration into various unsaturated soil textures [25] over long periods. Valuable features of this model are the relatively low computational cost and the large-scale applicability. These features are essential for integrating the hydrologic–hydrogeologic model into an integrated model for the identification of water related hazards as well as supporting an early warning system for the reduction of hydrogeological risk.

The Talbot–Ogden model is derived from the unsaturated Darcy's law and conservation of mass for water moving through a variably saturated porous media. It quickly simulates the infiltration in the water content-depth (θ-z) domain as its new perspective. In Figure 1, the fundamental idea in the Talbot–Ogden model is presented. In this model, bins are constructed by discretizing the moisture content domain as shown in Figure 1a. According to their water content values, they are independently arranged in parallel, not in series. Bins represent a collection of pore sizes corresponding to a specific range of moisture content $\theta_i \leq \theta \leq \theta_e$ in a soil. Within a particular bin, this range of moisture content can be found throughout the soil over the vertical domain (Figure 1b). Since the model has only one spatial dimension, this assumption is valid regardless of how sufficiently small the horizontal discretization is. The Green–Ampt equation is transformed and applied to compute the depth infiltration independently in each bin. A process called redistribution, which is invoked at every time step immediately after infiltration, governs the horizontal inter-bin flow along the θ-axis according to the capillary pressure associated with each bin. This process will take into account all the saturated bins but not only restricted to the local neighborhoods of different bins. The infiltration and redistribution are respectively driven by gravitational and relative capillary forces in each bin. During the infiltration, the capillary pressure and hydraulic conductivity become dynamic, and the wetting front as in the Green–Ampt model (Figure 1a) may not exist. The discretized water content domain has also been extended to be affine multi-dimensional [26] for depicting more complicated pore size distributions. It is an intrinsically mass conservative model that can be applied to various soil textures [8,23,27]. However, its suitability is directly related to the uncertain number of bins because the predicted flux is highly nonlinear with respect to the discretization in the moisture content domain. Therefore, this uncertainty plays an important role under different soil conditions in this model. The convergence test for choosing a proper number of bins by Talbot and Ogden [23] is more rigorously analyzed in this work, and its physical meaning, a greater variety of pore sizes leading to a larger infiltration rate, can be naturally explained from this study. It will also be quantitatively estimated how an infinite water content discretization affects the flux variation through an asymptotic analysis by linearly fitting the wetting front. It directly indicates that a particular depth ratio of the deepest to the shallowest bin fronts can maximize the infiltration flux.

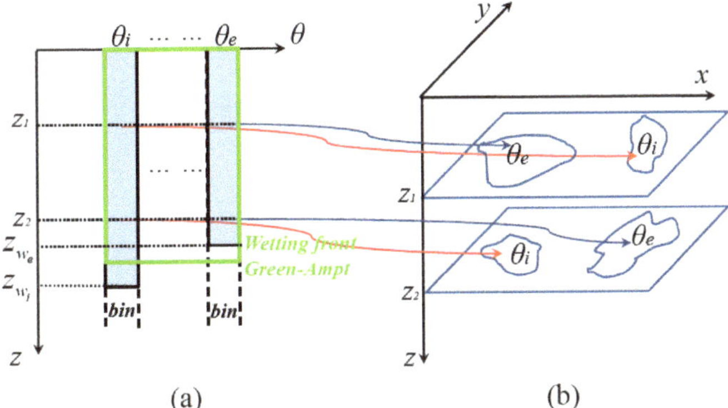

Figure 1. The mechanism in the Talbot–Ogden model. (**a**) Bins corresponding to different porosity θ-values compared to the wetting front in the Green–Ampt model; (**b**) Different moisture content in a soil over the vertical domain.

In Section 2, the Talbot–Ogden model is first introduced, which allows us to create a fast solver to simulate infiltration using finite water content discretization. A detailed sensitivity analysis is then made on the quantitative change of the flux as a function of the number of bins, and this work is generalized to an asymptotic analysis that gives an upper bound for infiltration flux variation. In Section 3, the infiltration flux variation using this model is firstly estimated from the physical parameters of a variety of real soil textures. Infiltration simulations for examples in both coarse and fine soil textures are presented using different numbers of bins to validate the theoretical analysis. Section 4 is the conclusive part.

2. Theory and Methodology

2.1. The Talbot–Ogden Model

Before presenting the Talbot–Ogden model, let us briefly review the Green–Ampt model relevant to it. The Green–Ampt equation [13], which is based on Darcy's law, provides a very simple model to describe the infiltration of water into the subsurface soil. By neglecting the depth of ponded water, the Green–Ampt equation for vertical infiltration is given by:

$$f = K_s\left(\frac{(\theta_d - \theta_i)H_c}{F(t)} + 1\right), \tag{1}$$

where f is the infiltration rate, K_s is saturated hydraulic conductivity, θ_i is initial moisture content, θ_d is the maximum moisture content during infiltration, H_c is the effective capillary drive at the wetting front, and $F(t)$ is the cumulative infiltration depth at time t.

The Talbot–Ogden model discretizes the entire water content domain into bins that flow in soils based on the porosity [23] and there is only one vertical spatial dimension (Figure 2) in this model. From the initial moisture content θ_i to the effective porosity θ_e there are n bins indexed by j with equal bin width $\Delta\theta$. The midpoint value $(j-1)\Delta\theta + \Delta\theta/2$ represented by θ_j is the moisture content of the j-th bin and the depth of its saturated wetting front is z_j. The j-th bin is assumed to be either fully saturated or dry at any depth. The residual moisture content is θ_r. The rightmost saturated bin is θ_d. The bins between θ_d and θ_e are unsaturated but can become saturated later. In the Green–Ampt model, the cumulative infiltration function $F(t)$ is defined as $F(t) = z(\theta_d - \theta_i)$. If we let $f = dF/dt = (\theta_d - \theta_i)dz/dt$, the vertical infiltration formula by substituting this expression into Equation (1) is obtained.

$$\frac{dz}{dt} = \frac{1}{(\theta_d - \theta_i)}\left(\frac{K_s H_c}{z} + K_s\right) \tag{2}$$

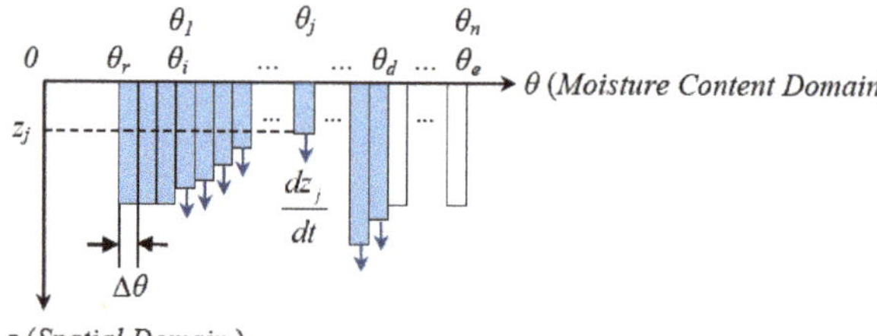

Figure 2. The infiltration in the Talbot–Ogden model.

Based on Equation (2), the front of the j-bin grows as:

$$\frac{dz_j}{dt} = \frac{1}{\theta_d - \theta_i}\left(\frac{K(\theta_d)\psi(\theta_d)}{z_j} + K(\theta_d)\right) \tag{3}$$

Here, ψ represents the capillary pressure.

The vertical infiltration of water in each bin is governed by Equation (3). The horizontal movement of water through bins is shown in Figure 3. By Equation (3), the infiltration rate dz_j/dt is inversely proportional to z_j if the other parameters are constant for the j-th bin. Hence, bins to the right tend to have greater front depths than the left ones especially in the beginning of infiltration (Figure 3a).

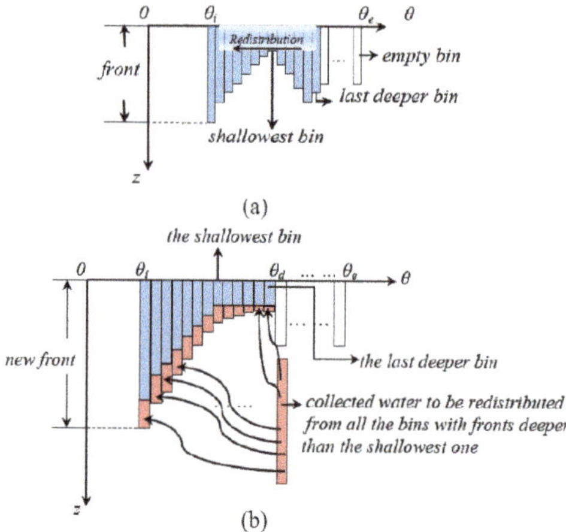

Figure 3. The redistribution in the Talbot–Ogden model: (**a**) pre-redistribution and (**b**) post-redistribution.

Due to the capillary effect in soil, water in bins with large θ-values tends to flow to those bins with smaller θ. This horizontal flow is referred to as the redistribution. It reorganizes the redundant water collected from those protruding wetting fronts in all saturated bins proportional to the values of the capillary pressure $\psi(\theta_j)$ of every bin participating in the redistribution [23]. In the redistribution process (Figure 3b), the last deeper bin is defined as the first saturated bin found over the moisture content domain along the negative direction of the θ-axis whose front depth is higher than the shallowest saturated bin. Note that the wetting front depths gradually decrease from left to right after the redistribution since the capillary pressure in the bins on the left acts immediately on water found at depth in the bins to the right in this model [23].

The reasons for choosing a fixed moisture content θ_d for both $K(\theta)$ and $\psi(\theta)$ in Equation (3) are found in [23]. Water tends move downward through the saturated bins with large θ-values, which is the reason that $K(\theta_d)$ is chosen for every bin. However, the capillary pressure in a soil with only lower moisture content than the current saturated bin is always satisfied prior to the rightmost saturated bin, which is the reason that $\psi(\theta_d)$ is chosen for every bin. The functions $K(\theta)$ and $\psi(\theta)$ are from Brooks and Corey [28], but other soil hydraulic models [29–31] can be used without affecting the analysis and conclusions in this paper.

Due to the uncertain discretization of the moisture content domain into n bins, a dynamical system [32] is generated

$$z_j^{t+\Delta t} = z_j^t + \frac{dz_j^t}{dt} \times \Delta t + redist_j^t, j = 1\ldots n, t \in [0, T],\tag{4}$$

and

$$redist_j^t = redist_j^t\left(\psi(\theta_1), z_1^t, \psi(\theta_2), z_2^t, \ldots, \psi\left(\theta_{last\ deeper\ bin}\right), z_{last\ deeper\ bin}^t\right)\tag{5}$$

In Equations (4) and (5), the term redist represents the redistribution as the inter-bin movement of water.

The second term of the right hand side in Equation (4) comes from Equation (3) as infiltration. The computation is inexpensive using Equation (4), as for every time step only n ordinary different equations are solved for the infiltration. The simulation of redistribution takes $O(n)$ operations for all n bins. In the numerical convergence test by Talbot and Ogden, the largest n is 400 with a time step size $2.5\ s$ for the coarsest soil sand, which provides a fast, accurate solution.

2.2. Instantaneous Infiltration Rates Analysis

The most important quantity in the Talbot–Ogden model is the bin width $\Delta\theta$. Therefore, the range of $\Delta\theta$ must be selected to match the unsaturated flow in specific soil systems because the flux is nonlinear with respect to $\Delta\theta$. There are two simple assumptions made before analyzing the flux: (1). θ_d is assumed to be fixed and independent of the number of bins, which is reasonable since θ_d corresponds to the rightmost saturated bin that is changing during the infiltration; (2). All bins with $\theta_i \le \theta \le \theta_d$ are already saturated. If there are empty bins, it means that the surface water can be absorbed in the next time step so that the instantaneous infiltration rate equals the precipitation rate.

The number of bins and its nonlinear influence on the predicted flux can be considered under these assumptions. Now the only uncertain quantity in the Talbot–Ogden model is the number of bins, which means either a finer or a coarser discretization. All other parameters remain constant. The unsaturated flow divides into two steps within every time step: infiltration and redistribution (Figure 4).

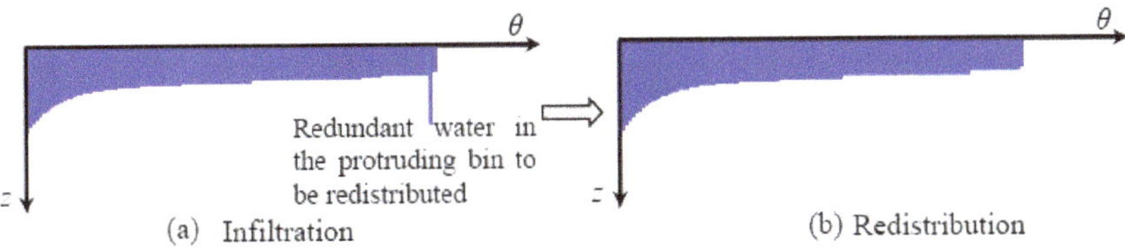

Figure 4. A numerical example: (**a**) infiltration and (**b**) redistribution.

The infiltration governs the vertical downward movement of water in every bin. The redistribution governs the horizontal inter-bin flow. The amount of water redistributed to each bin is proportional to the local capillary pressure. In Figure 4, redundant water in the protruding bin caused by infiltration is redistributed to the bins to its left. In the next time step, the process begins with another infiltration followed by redistribution (Equation (4)). These two processes alternate until the simulation completes. The strength of the peak in that particular protruding bin with more infiltration will depend on the bin width $\Delta\theta$ due to its nonlinearity effect in Equation (4). The redistribution phase moves no surface water downward into the soil by assumption, but only rearranges the wetting fronts in different bins. Due to the properties of the decreasing front depths by the redistribution, an analysis of the instantaneous infiltration rates in the Talbot–Ogden model can be made.

2.2.1. One Bin versus Two Bins

Figure 5 presents the model of one bin and its division into two bins. The rectangle *ACHF* represents bin_1. Its wetting front is *FH*, its width is $\Delta\theta_1$, and its depth is z_1. Similarly, the two bins in Figure 5 are bin_2 (*ABJI*) and bin_3 (*BCED*) with wetting fronts *IJ* and *DE*, respectively. Both their bin widths are equal to $\Delta\theta_2$, which means $\Delta\theta_1 = 2\Delta\theta_2$. Their front depths are z_2 and z_3. It is assumed that the entire water content for the two cases is the same. So, the water in bin_1 equals that contained in the union of bin_2 and bin_3, that is $z_1 \times \Delta\theta_1 = z_2 \times \Delta\theta_2 + z_3 \times \Delta\theta_2$. These two cases can be compared. Let V_{Onebin} and $V_{Twobins}$ denote the instantaneous infiltration rates for each case. V_{Onebin} is calculated by

$$
\begin{aligned}
V_{Onebin} &= V_{bin_1} = \frac{dz_1}{dt} \times \Delta\theta_1 \\
&= \frac{2 \times \Delta\theta_2 \times K(\theta_d)}{\theta_d - \theta_i}\left(\frac{\psi(\theta_d)}{z_1} + 1\right)
\end{aligned}
\tag{6}
$$

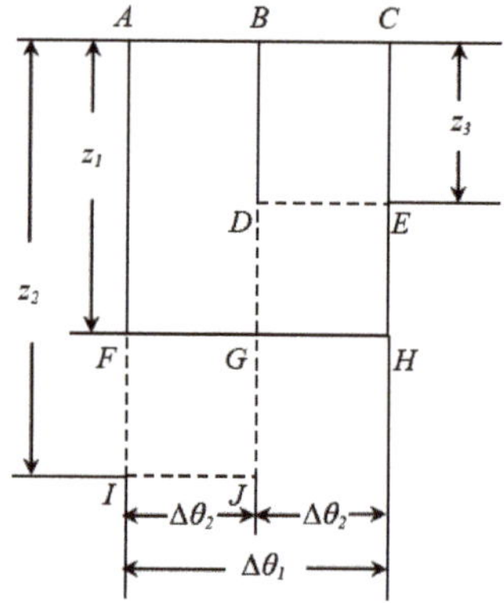

Figure 5. One bin versus two bins.

$V_{Twobins}$ is calculated by

$$
\begin{aligned}
V_{Twobins} &= V_{bin_2} + V_{bin_3} = \left(\frac{dz_2}{dt} + \frac{dz_3}{dt}\right) \times \Delta\theta_2 \\
&= \frac{\Delta\theta_2 K(\theta_d)}{\theta_d - \theta_i}\left(\frac{\psi(\theta_d)}{z_2} + \frac{\psi(\theta_d)}{z_3} + 2\right)
\end{aligned}
\tag{7}
$$

Based on the capillary pressure of the soil, a relationship among depths exists: $z_2 > z_1 > z_3$, since left bins always have deeper wetting fronts than the bins to their right. Moreover, if $z_2 = z_3$, then splitting one bin into two bins reverts back to the one bin case. Now, the difference between the instantaneous infiltration rates can be computed by subtracting (6) from (7). Thus,

$$
\begin{aligned}
V_{Twobins} - V_{Onebin} &= V_{bin_2} + V_{bin_3} - V_{bin_1} \\
&= \frac{\Delta\theta_2 \times K(\theta_d) \times \psi(\theta_d)}{\theta_d - \theta_i}\left(\frac{(z_2 - z_3)^2}{z_2 z_3 (z_2 + z_3)}\right) \\
&\geq 0 \ (with\ equality\ if\ z_2 = z_3).
\end{aligned}
\tag{8}
$$

Equation (8) uses $z_1 \times \Delta\theta_1 = z_2 \times \Delta\theta_2 + z_3 \times \Delta\theta_2$. Therefore, when one bin is split into two bins and all possible pores are saturated, additional water infiltrates into the soil faster. Its increment can be bounded by (8). Doubling the number of bins leads to a more continuous flow among bins. Hence, more bins mean a higher infiltration rate. One bin is increased to n bins to prove this assumption.

2.2.2. One Bin versus n Bins

Similarly, n bins can be obtained by equally discretizing the moisture content domain of the one bin case into n pieces. All other parameters remain constant. In Figure 6, the rectangle $ABCD$ represents one bin bin_X. Its depth and bin width are z_X and $\Delta\theta_X$, respectively. Then bin_X is split into n bins, marked by $bin_1, bin_2, \dots, bin_{n-1}$, and bin_n. All these new bins have the same width $\Delta\theta$. Their depths from left to right are $z_1, z_2, \dots, z_{n-1}$ and z_n. It is similar to the previous example to have $z_X \times \Delta\theta_X = (z_1 + z_2 + \dots + z_n) \times \Delta\theta$, $n\Delta\theta = \Delta\theta_X = \theta_d - \theta_i$, and $z_1 \geq z_2 \geq \dots \geq z_n$. By geometry there exists an index l such that $z_1 \geq z_2 \geq \dots \geq z_l \geq z_X \geq z_{l+1} \geq \dots \geq z_n$. The infiltration rate of bin_X is

$$
\begin{aligned}
V_{Onebin} &= V_{bin_X} = \frac{dz_X}{dt} \times \Delta\theta_X \\
&= \frac{\Delta\theta_X K(\theta_d)}{\theta_d - \theta_i}\left(\frac{\psi(\theta_d)}{z_X} + 1\right)
\end{aligned}
\tag{9}
$$

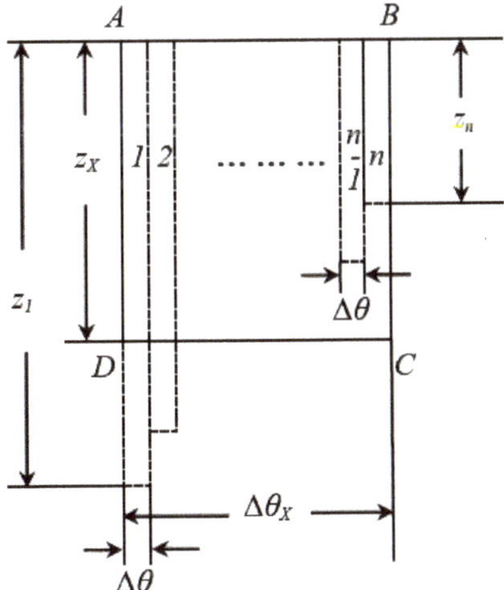

Figure 6. One bin versus n bins.

For the *n-bin* case, the summed infiltration rate is

$$
\begin{aligned}
V_{nbins} &= \sum_{j=1}^{n} V_{bin_j} = \sum_{j=1}^{n} \frac{dz_j}{dt} \times \Delta\theta \\
&= \sum_{j=1}^{n} \frac{\Delta\theta K(\theta_d)}{\theta_d - \theta_i}\left(\frac{\psi(\theta_d)}{z_j} + 1\right)
\end{aligned}
\tag{10}
$$

By Equation (9) and Equation (10), the infiltration difference between these two cases is

$$
\begin{aligned}
V_{nbins} - V_{Onebin} &= \left(\sum_{j=1}^{n} V_{bin_j} \right) - V_{bin_X} \\
&= \frac{\Delta\theta \times K(\theta_d) \times \psi(\theta_d)}{\theta_d - \theta_i} \left(\sum_{j=1}^{n} \frac{1}{z_j} - \frac{n}{z_X} \right) \\
&= \frac{\Delta\theta \times K(\theta_d) \times \psi(\theta_d)}{\theta_d - \theta_i} \left(\sum_{j=1}^{n} \frac{1}{z_j} - \frac{n^2}{\sum_{j=1}^{n} z_j} \right) \geq 0
\end{aligned}
\tag{11}
$$

Note that the last step of Equation (11) uses

$$
\frac{\sum_{j=1}^{n} z_j}{n} \geq \left(\prod_{j=1}^{n} z_j \right)^{\frac{1}{n}} \geq \frac{n}{\sum_{j=1}^{n} \frac{1}{z_j}}.
\tag{12}
$$

Therefore from Equation (11), which means $V_{nbins} - V_{Onebin} \geq 0$, more bins in the model result in a greater infiltration rate.

Now, n_1 bins is increased to n_2 bins, where n_2 and n_1 are integers and $n_2 > n_1$. This case resembles the extension from one bin to $\lceil n_2/n_1 \rceil$ bins. The difference in infiltration rates between these n_1 and n_2 bins is bounded by $V_{n_2 bins} - V_{n_1 bins} \approx V_{\lceil n_2/n_1 \rceil bins} - V_{Onebin}$. By Equation (11), we can conclude that if a soil texture can be fitted by a finer discretization, its overall conductivity becomes higher in the Talbot–Ogden model. We found the infiltration rate using asymptotic analysis.

2.3. Asymptotic Analyses and Its Physical Meaning

Asymptotic analysis is used when the number of bins increases toward infinity. Equation (11) and the assumption $n\Delta\theta = \theta_d - \theta_i$ are used first:

$$
V_{nbins} - V_{Onebin} = K(\theta_d) \times \psi(\theta_d) \left(\frac{1}{n} \sum_{j=1}^{n} \frac{1}{z_j} - \frac{n}{\sum_{j=1}^{n} z_j} \right)
\tag{13}
$$

Before asymptotic analysis, the wetting front curve is loosely assumed to approximate a straight slanted line (Figure 7) in the θ-z domain. This assumption can also be demonstrated by the water content profiles in the solutions of the Richards Equation using Hydrus-1D [27] under many circumstances. Although it is only an approximation, all the wetting fronts in this model decrease from z_1 to z_n (Figure 6) after redistribution in every time step. When $n \to \infty$, the effect of the number of bins on the instantaneous infiltration rate in the Talbot–Ogden model is:

$$
\begin{aligned}
&\lim_{n\to\infty} \left(V_{nbins} - V_{Onebin} \right) \\
&= \lim_{n\to\infty} \left[K(\theta_d) \times \psi(\theta_d) \times \left(\frac{(z_1 - z_n) \sum_{j=1}^{n} \frac{1}{z_j}}{n(z_1 - z_n)} - \frac{z_1 - z_n}{\frac{z_1 - z_n}{n} \sum_{j=1}^{n} z_j} \right) \right] \\
&= K(\theta_d) \times \psi(\theta_d) \times \left(\frac{1}{z_1 - z_n} \int_{z_n}^{z_1} \frac{1}{z} dz - \frac{z_1 - z_n}{\int_{z_n}^{z_1} z dz} \right) \\
&= K(\theta_d) \times \psi(\theta_d) \times \left(\frac{1}{z_1 - z_n} \ln\frac{z_1}{z_n} - \frac{2}{z_1 + z_n} \right).
\end{aligned}
\tag{14}
$$

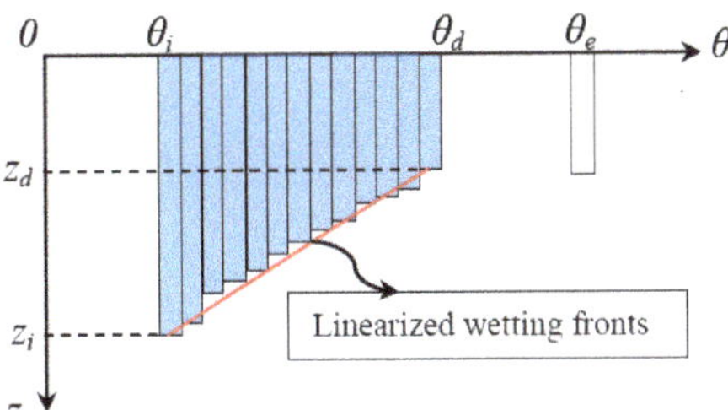

Figure 7. Approximately linearly decaying wetting fronts.

Equation (14) is obtained by the fundamental law of integration, which is the Newton–Leibniz rule. Summing the reciprocals and then taking the average results in a positive number. Therefore, in the integration the lower bound is $1/z_n$, and the upper bound is $1/z_1$.

Moreover, z_1 corresponds to the deepest saturated bin, whereas z_n corresponds to the shallowest one. For convenience let z_i and z_d denote these two depths, respectively, because of the moisture content they represent. Thus, when $n \to \infty$,

$$
\begin{aligned}
\|V_{nbins} - V_{Onebin}\|_{L_1} &= \lim_{n \to \infty} (V_{nbins} - V_{Onebin}) \\
&= K(\theta_d) \times \psi(\theta_d) \times \left(\frac{1}{z_i - z_d} ln \frac{z_i}{z_d} - \frac{2}{z_i + z_d} \right)
\end{aligned}
\tag{15}
$$

What Equation (15) tells us is that the variation in infiltration rate depends only on the depths corresponding to the initial porosity θ_i and the moisture content θ_d of the rightmost saturated bin. These two special bins are actually the deepest and the shallowest ones, respectively. An interpretation is that the whole wetting front is pushed downward by water in the largest saturated porosity θ_d and is prevented from progressing by water in the initial moisture content θ_i. Therefore, the advancement of the wetting front is a compromise between these two bins. Interestingly, the unit of length does not count in the expression, and the quantity $\frac{1}{z_i - z_d} ln \frac{z_i}{z_d} - \frac{2}{z_i + z_d}$ is dimensional and is meaningful in physics. However, its value needs some further research, which is crucial for this model. The following practical case is presented: suppose the two wetting fronts satisfy

$$
\exists \delta > 0 \text{ and } \hat{N} \in \mathbb{Z}^+, \text{ such that } z_d(t) + \hat{N}\delta \geq z_i(t) \geq z_d(t) + \delta, \quad \text{for all } t.
$$

If $\lim_{t \to \infty} z_d(t) \to \infty$, then the following limits hold:

$$
\lim_{z_d \to \infty} \left(\frac{1}{z_i - z_d} ln \frac{z_i}{z_d} - \frac{2}{z_i + z_d} \right) \leq \lim_{z_d \to \infty} \left(\frac{1}{\delta} ln \left(1 + \frac{\hat{N}\delta}{z_d} \right) - \frac{2}{2z_d + \hat{N}\delta} \right) = 0
\tag{16}
$$

and

$$
\lim_{z_d \to \infty} \left(\frac{1}{z_i - z_d} ln \frac{z_i}{z_d} - \frac{2}{z_i + z_d} \right) \geq \lim_{z_d \to \infty} \left(\frac{1}{\hat{N}\delta} ln \left(1 + \frac{\delta}{z_d} \right) - \frac{2}{2z_d + \delta} \right) = 0.
\tag{17}
$$

Therefore, it follows that

$$
\lim_{z_d \to \infty} \left(\frac{1}{z_i - z_d} ln \frac{z_i}{z_d} - \frac{2}{z_i + z_d} \right) = 0.
\tag{18}
$$

Equation (18) means that if the wetting fronts corresponding to the initial moisture content θ_i and the effective porosity θ_e are within some distance from each other in depth, the infiltration in the Talbot–Ogden model will eventually become a steady state flow. However, a transformation will make this discussion easier.

Without loss of generality, let $z_i = r z_d$ with $r > 1$, then a function $D(r)$ characterizing the difference between V_{nbins} and V_{Onebin}, is defined by $D(r) = ln(r)/(r - 1) - 2/(r + 1)$. It follows that

$$\|V_{nbins} - V_{Onebin}\|_{L_1} = \frac{K(\theta_d) \times \psi(\theta_d)}{z_d} \times D(r). \tag{19}$$

The potential maximum or minimum of $D(r)$ can be attained by taking its derivative

$$\frac{dD(r)}{dr} = -\frac{ln(r)}{(r+1)^2} + \frac{1}{r(r-1)} + \frac{2}{(r+1)^2}, \tag{20}$$

which has a zero point at $r \approx 8.16$. There are also two limits for two extreme cases:

$$\lim_{r \to 1^+} D(r) = \lim_{r \to 1^+} \frac{(r+1)ln(r) - 2(r-1)}{r^2 - 1} = 0, \tag{21}$$

and

$$\lim_{r \to +\infty} D(r) = \lim_{r \to +\infty} \left[\frac{1}{r-1} ln(r) - \frac{2}{r+1} \right] = 0. \tag{22}$$

Equation (21) corresponds to the Green–Ampt model. Equation (22) means that if the deepest wetting front is too far away from the shallowest one, then the infiltration rate also tends to the one bin case (Figure 8).

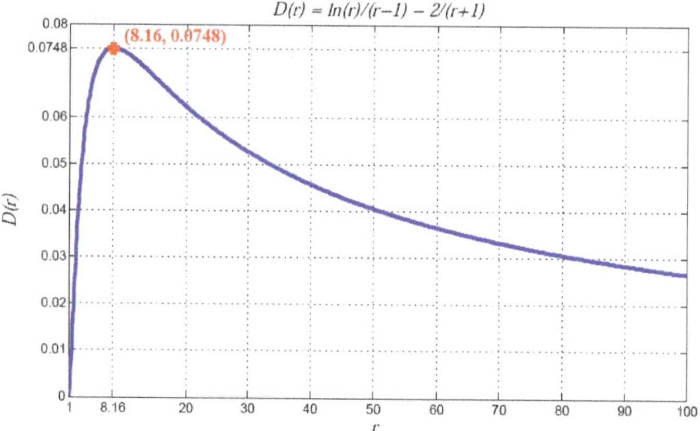

Figure 8. Ratio r of the deepest wetting front to the shallowest one and its influence on $D(r)$.

The curve in Figure 8 very clearly shows the change in $D(r)$. With the maximum value of $D(r)$, it is able to bound $\|V_{nbins} - V_{Onebin}\|_{L_1}$. Note that

$$\|V_{nbins} - V_{Onebin}\|_{L_1} = \frac{K(\theta_d) \times \psi(\theta_d)}{z_d} \times D(r) \leq \frac{0.0748 \times K_s \times \psi_b \times \left(\frac{\theta_d - \theta_i}{\theta_e - \theta_i} \right)^{3 + \frac{1}{\lambda}}}{z_d} \ (by\ Brooks\ and\ Corey) \tag{23}$$

where λ denotes the pore size distribution index [14]. If $\theta_d = \theta_e$, then the hydraulic models used in inequality (23) are unimportant because $K(\theta_d) = K_s$ and $\psi(\theta_d) = \psi_b$ always hold. In this situation, hence,

$$\|V_{nbins} - V_{Onebin}\|_{L_1} \leq 0.0748 \times \frac{K_s \times \psi_b}{z_d}, \tag{24}$$

the upper bound of the infiltration rate error can be calculated. Note that if a soil texture is chosen, then z_d can grow as a function of time and the rainfall rate, therefore making this upper bound decrease with respect to time.

3. Numerical Experiments

In this section, the flux variation upper bound is calculated from the physical parameters of a variety of soil textures. Numerical tests are then carried out to simulate infiltration and verify our analysis in the previous section.

If z_d is fixed in Equation (24), what is important is the product of saturated hydraulic conductivity K_s and the bubbling pressure ψ_b. Therefore, $0.0748 \cdot K_s \psi_b$ gives a bound for discrepancy of infiltration rates. In Table 1 [23], $K_s \psi_b$ decreases from coarser to finer soils. The one exception is between clay loam and silty clay loam. In general, the upper bound of $\|V_{nbins} - V_{Onebin}\|_{L_1}$ becomes larger with coarser soils. Significantly, this finding means that if the soil is finer, then the infiltration will not be distinguishable based on a change in the number of bins. However, in coarse textured soil systems, the number of bins is more important than in finer ones. This is the reason why more bins may be required to test the coarser soil: the outcome varies in a wider range.

Table 1. Parameters of different soil systems.

Texture	K_s (cm/h)	ψ_b (cm)	$K_s \psi_b$ (cm²/h)	$0.0748 \cdot K_s \psi_b$ (cm/h)
Sand	23.56	7.26	171.05	12.795
Loamy sand	5.98	8.69	51.18	3.828
Sandy loam	2.18	14.66	31.96	2.391
Loam	1.32	11.15	14.72	1.101
Silt loam	0.68	20.79	14.14	1.058
Sandy clay loam	0.30	28.08	8.42	0.630
Clay loam	0.20	25.89	5.17	0.386
Silty clay loam	0.20	32.56	6.51	0.487
Sandy clay	0.12	29.17	3.50	0.262
Silt clay	0.10	34.19	3.42	0.256
Clay	0.06	37.30	2.24	0.168

From the last column of Table 1, if z_d is fixed and not too deep nor too shallow, then the value $0.0748 \cdot K_s \psi_b$ is in fact a certain bound for the variation in infiltration rate. This condition means that if the rainfall rate is around that value or of the same scale for a specific soil system, then the choice of a proper number of bins to make the simulation realistic is needed. Hence, the choice of the number of bins should be determined at least by both the soil and rainfall rates. Note that when z_d is very large in inequality (24), the bound becomes small, which means the Talbot–Ogden model acts as the Green–Ampt model or the Richards model for the steady state flow.

Using the Talbot–Ogden method it is possible to simulate this infiltration process by choosing some parameters, including an appropriate number of bins. If the proper number of bins is relatively large, then the modeled soil lets water quickly pass through it, which indicates that an underground reservoir recharges quicker. On the other hand, if fewer bins are needed, then the model indicates that this soil has a good capacity to retain water, which is beneficial for growing plants and for agriculture. In general, the number of bins may indicate the pore size categories that can be modeled in the soil for infiltration.

The comparison of instantaneous infiltration rates computed by Talbot–Ogden method and Hydrus-1D is tested. In this work, the main focus is the influence of the number of bins on the infiltration rates in the Talbot–Ogden method. The soil parameters used are listed in Table 2 [14]. The time step Δt is set as 10 s for all the simulations to guarantee the comparison consistency. Figure 9 shows the infiltration rate curves corresponding to different numbers of bins for three soil textures. These three types are sand, sandy clay, and silt loam. Different rainfall rates are used because each soil type can generate comparable data only under the proper rainfall rate. One continuous rainfall is one pulse. There are two pulses, and each lasts for 1.5 h. The numbers of bins used in testing are 25, 125, and 250.

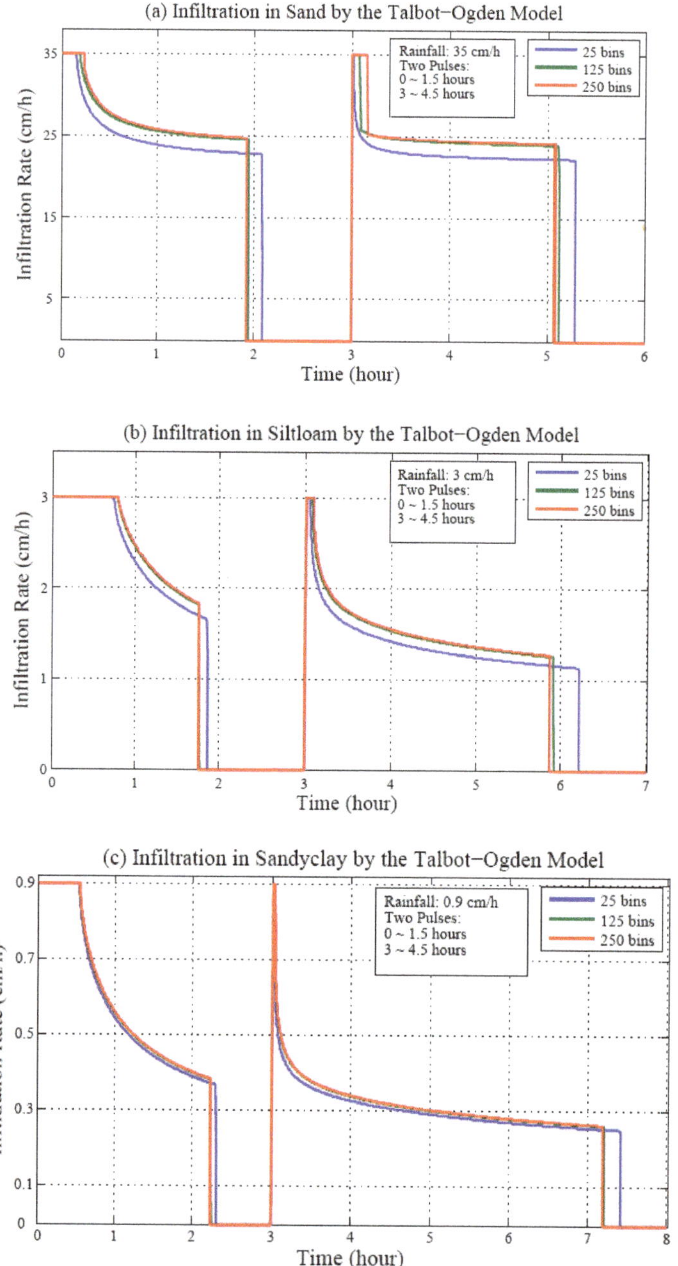

Figure 9. Comparison of infiltration rates in (**a**) sand, (**b**) silt loam, and (**c**) sandy clay using different numbers of bins.

Table 2. Soil textures and parameters used in model evaluation.

Soil	K_s (cm/h)	ψ_b (cm)	θ_r	θ_i	θ_e	λ
Sandy clay	0.12	29.17	0.109	0.239	0.321	0.223
Silt loam	0.68	20.79	0.015	0.133	0.486	0.234
Sand	23.56	7.26	0.02	0.033	0.417	0.694

With an increase in the numbers of bins, the infiltration rate in any soil system rises. During the second precipitation, this phenomenon is even more obvious than in the first one. The reason is explained by the analysis given in Section 2. At the beginning of the second rainfall there are already some dry pores with larger pores that correspond to θ near θ_e, which is to the right of the moisture content domain. The depth z_d is very small at this moment so that the change in infiltration rate is more sensitive to the perturbation of the numbers of bins than that in the first pulse. Mathematically, if $z_d < 1$ cm (the unit is consistent with that in Table 1), the upper bounds given by inequality (24) and Table 1 are relatively easier to approach during the second pulse, thus verifying the analysis in Section 2.

The quadratic or root mean squared (*RMS*) difference in the instantaneous infiltration rates calculated at every time step with 25 bins and 125 or 250 bins, are defined as

$$RMS_1 = \sqrt{\frac{\sum_{time\ step=1}^{N_1}\left(f_{25\ bins}^{time\ step}-f_{125\ bins}^{time\ step}\right)^2}{N_1}} \tag{25}$$

and

$$RMS_2 = \sqrt{\frac{\sum_{time\ step=1}^{N_2}\left(f_{25\ bins}^{time\ step}-f_{250\ bins}^{time\ step}\right)^2}{N_2}}, \tag{26}$$

where f represents the instantaneous infiltration rates as computed for different numbers of bins, respectively, and N is the number of f values in the interval for which *RMS* is calculated. In Table 3, the columns of *RMS* values and their comparisons show that the effect on the infiltration rate becomes smaller with an increase in the number of bins. In the fourth column of Table 3, the decreasing RMS_1/RMS_2 values indicate that the coarser the soil texture is, the more sensitive the infiltration rate is to the change in the number of bins, especially at the beginning of this change. This phenomenon can be attributed to the relatively large range of infiltration change for coarser soils as shown in Table 1. Note that the root mean square values in Table 3 also satisfy $RMS_{1,2} < 0.0748 \cdot K_s \psi_b$.

Table 3. Influence of the number of bins on the infiltration in the Talbot–Ogden model.

Type	RMS_1 (cm/h)	RMS_2 (cm/h)	RMS_1/RMS_2 (%)	$0.0748 \cdot K_s \psi_b$ (cm/h)
Sandy Clay	0.0171	0.0182	94.11	0.262
Silt Loam	0.1256	0.1515	82.91	1.058
Sand	1.9330	2.6853	71.99	12.795

Root mean sqaure$_1$ = The difference of infiltration rates between 25 and 125 bins; Root mean square$_2$ = The difference of infiltration rates between 25 and 250 bins.

Figure 10a shows how infiltration behaves when the number of bins increases rapidly. There are three arrays of data for comparison: one with 100 bins, another with 500 bins, and the rest with bins. The rainfall rate is set to 2 cm/h, and two pulses are selected with each lasting 1.5 h. Figure 10b shows similar results but with a coarser soil system silt loam. Let N denote the number of time steps where infiltration discrepancies appear corresponding to different numbers of bins, the following equation is used to calculate the influence of the rapid increase in the number of bins:

$$Influence\ of\ large\ number\ of\ bins(x) = \sqrt{\frac{\sum\limits_{i=1}^{N}\left(\frac{V_{i,100\ bin}-V_{i,x\ bin}}{V_{i,100\ bin}}\right)^2}{N}} \times 100\%. \tag{27}$$

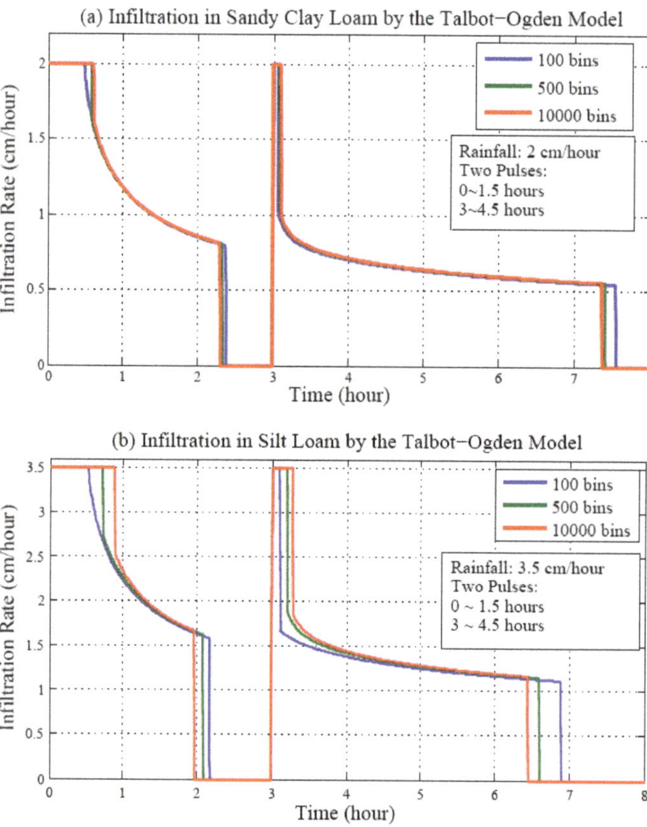

Figure 10. Infiltration in (**a**) sandy clay loam, and (**b**) silt loam with a large increase in the number of bins.

For the three cases in Figure 10a,b, the use of (27) yields Table 4. In Table 4, for sand clay loam, the difference in average infiltration rates using 100 bins and bins is

$$2.0\ cm/h \times 8.56\% = 0.1712\ cm/h < 0.0748 \cdot K_s\psi_b = 0.630\ cm/h. \tag{28}$$

Table 4. Influence of large number of bins on the infiltration rate compared with 100 bins as the baseline.

Number of Bins (x)	Sand Clay Loam	Silt Loam
500	8.04%	17.48%
	8.56%	24.32%

Similarly, for silt loam, their difference is

$$3.5\ cm/h \times 24.32\% = 0.8512\ cm/h < 0.0748 \cdot K_s\psi_b = 1.058\ cm/h. \tag{29}$$

In Equations (28) and (29) show that if the number of bins increases by 100 times, then the increment in the infiltration rates is still bounded by the data in Table 1. In fact, the increment of infiltration rate is more significant from 100 bins to 500 bins. These results show that infiltration in the finer soil system is less sensitive to the rapid increase in the number of bins than the coarser one, which is consistent with the analysis in Equation (24) and Table 1.

4. Conclusions

A sensitivity analysis of the instantaneous infiltration rates in the Talbot–Ogden model is developed with rigorous mathematical deduction and reasonable physical assumptions. This analysis starts from the two-bin case to the n-bin case. It is concluded that the infiltration rate increases with finer discretization in the moisture content domain. Numerical experiments on the overall infiltration rates have confirmed this theoretical conclusion. Therefore, the choice of the number of bins is very important for different soil textures. When θ_d approaches θ_e, the Talbot–Ogden model always generates higher infiltration fluxes than the Green–Ampt model, where $K = K_s$ and $H_c = -\psi_b$ are set.

An asymptotic analysis is also made on estimating the largest infiltration rates. Using the loose assumption of a line-shaped wetting front, the asymptotic analysis provides an upper bound of the infiltration rate difference resulting from two arbitrarily different moisture content discretizations. The numerical experiments then illustrate this result. These upper bounds for different soil textures can determine the accuracy of the Talbot–Ogden model in predicting the fluxes in various environments. Note that if the wetting front is apparently nonlinear, which is possible in the Talbot–Ogden model especially when precipitations are intermittent, then this asymptotic analysis may be adjusted to be piecewise for different intervals of the moisture content domain. The nonlinearity can be handled using exponential functions (i.e., z^α, $\alpha > 1$) to approximate the depths of the wetting fronts.

There are basically two factors determining the infiltration rates in the Talbot–Ogden model. One is the number of bins by the n-bin case analysis and the other is the front depth discrepancy between the leftmost bin and the rightmost bin in our asymptotic analysis. Moreover, it is intrinsically mass conservative and is always computable.

The analysis will be extended to predicting the infiltration rate variation as a function of time but not only restricted to the instantaneous moment. Meanwhile, bugs and cracks can be easily incorporated into this model, although this will make the flux analysis more complicated. This model can also be extended to heterogeneous soil textures by solving stacked homogeneous soil layers, which is future work.

Author Contributions: In this study, L.L. has made contributions to conceptualization, formal analysis, data curation, investigation and writing-review and editing. H.Y. has made his contributions to methodology, software, formal analysis, resources, and writing-original draft. Both authors have acquired funds for this research.

Funding: This research was sponsored by the Natural Science Foundation for Young Scientists of Jiangsu Province (grant no. BK20180450), the Natural Science Foundation of Nanjing University of Posts and Telecommunications NUPTSF (grant no. NY219080) and the China Postdoctoral Science Foundation (grant no. 2018T110531).

Acknowledgments: The authors gratefully acknowledge the assistance of Talbot, Ogden and Craig Douglas for their extraordinary insights and comments in the development of this paper. The authors also thank King Abdullah University of Science and Technology for the support on computational resources.

Conflicts of Interest: The authors declare no conflict of interest.

Physical Notations and Units

Variable	Physical Meaning	Unit
K_s	Saturated hydraulic conductivity	$[\mathrm{LT}^{-1}]$
K	Hydraulic conductivity	$[\mathrm{LT}^{-1}]$
ψ_b	Bubbling pressure	$[\mathrm{L}]$
ψ	Capillary pressure	$[\mathrm{L}]$
H_c	Suction head/Effective capillary pressure	$[\mathrm{L}]$
θ	Moisture content or porosity	$[\mathrm{L}^3\mathrm{L}^{-3}]$
θ_i	Initial moisture content	$[\mathrm{L}^3\mathrm{L}^{-3}]$
θ_d	The maximum moisture content during infiltration	$[\mathrm{L}^3\mathrm{L}^{-3}]$
θ_e	Effective moisture content	$[\mathrm{L}^3\mathrm{L}^{-3}]$
θ_r	The residual water content	$[\mathrm{L}^3\mathrm{L}^{-3}]$
$\Delta\theta$	Bin width after discretization	$[\mathrm{L}^3\mathrm{L}^{-3}]$
f, V	Infiltration rate	$[\mathrm{LT}^{-1}]$
F	Total water in soil	$[\mathrm{L}]$
t	Time	$[\mathrm{T}]$
z	Infiltration depth	$[\mathrm{L}]$
z_d	Front depth of the bin associated with θ_d	$[\mathrm{L}]$
z_{w_e}	Front depth of the bin associated with θ_e	$[\mathrm{L}]$
z_{w_i}	Front depth of the bin associated with θ_i	$[\mathrm{L}]$

References

1. De Vries, J.J.; Simmers, I. Groundwater recharge: An overview of processes and challenges. *Hydrogeol. J.* **2002**, *10*, 5–17. [CrossRef]
2. Sanford, W. Recharge and groundwater models: An overview. *Hydrogeol. J.* **2002**, *10*, 110–120. [CrossRef]
3. Smith, R.P. Geologic setting of the Snake River plain aquifer and vadose zone. *Vadose Zone J.* **2004**, *3*, 47–58. [CrossRef]
4. McElroy, D.L.; Hubbell, J.M. Evaluation of the conceptual flow model for a deep vadose zone system using advanced Tensiometers. *Vadose Zone J.* **2004**, *3*, 170–182. [CrossRef]
5. Wilson, L.G. Monitoring in the vadose zone part I: Storage changes. *Gr. Water Monit. Remediat.* **1981**, *1*, 32–41. [CrossRef]
6. Charbeneau, R.J. *Groundwater Hydraulics and Pollutant Transport*; Waveland Press: Long Grove, IL, USA, 2006.
7. Manenti, S.; Amicarelli, A.; Todeschini, S. WCSPH with limiting viscosity for modeling landslide hazard at the slopes of artificial reservoir. *Water* **2018**, *10*, 515. [CrossRef]
8. Sophecleous, M. Interactions between groundwater and surface water: The state of the science. *Hydrogeol. J.* **2002**, *10*, 52–67. [CrossRef]
9. Huang, J.; Liao, W.; Li, Z. A multi-block finite difference method for seismic wave equation in auxiliary coordinate system with irregular fluid–solid interface. *Eng. Comput.* **2018**, *35*, 334–362. [CrossRef]
10. Richards, L.A. Capillary conduction of liquids in porous mediums. *Physics* **1931**, *1*, 318–333. [CrossRef]
11. Celia, M.A.; Bouloutas, E.T. A general mass-conservative numerical solution for the unsaturated flow equation. *Water Resour. Res.* **1990**, *26*, 1483–1496. [CrossRef]
12. Schneid, E.; Knabner, P.; Radu, F. A priori error estimates for a mixed finite element discretization of the Richards' equation. *Numer. Math.* **2004**, *98*, 353–370. [CrossRef]
13. Green, W.H.; Ampt, C.A. Studies on soil physics. *J. Agric. Sci.* **1911**, *4*, 1–24. [CrossRef]
14. Rawls, W.J.; Asce, M.; Brakensiek, D.L.; Miller, N. Green–Ampt infiltration parameters from soils data. *J. Hydraul. Eng.* **1983**, *109*, 62–70. [CrossRef]
15. Huyakorn, P.S.; Pinder, G.F. *Computational Methods in Subsurface Flow*; Academic Press: New York, NY, USA, 1983.
16. Haverkamp, R.; Vauclin, M.; Touma, J.; Wierenga, P.J.; Vachaud, G. A comparison of numerical simulation models for one-dimensional infiltration. *Soil Sci. Soc. Am. J.* **1977**, *41*, 285–294. [CrossRef]

17. Grimaldi, S.; Petroselli, A.; Romano, N. Green–Ampt curve-number mixed procedure as an empirical tool for rainfall-runoff modeling in small and ungauged basins. *Hydrol. Process.* **2012**. [CrossRef]

18. Freyberg, D.L.; Reeder, J.W.; Franzini, J.B.; Remson, I. Application of the Green-Ampt Model to infiltration under time-dependent surface water depths. *Water Resour. Res.* **1980**, *16*, 517–528. [CrossRef]

19. Zadeh, K.S.; Shirmohammadi, A.; Montas, H.J.; Felton, G. Evaluation of infiltration models in contaminated landscape. *J. Environ. Sci. Health Part A* **2007**, *42*, 983–988. [CrossRef] [PubMed]

20. Voller, V.R. On a fractional derivative form of the Green–Ampt infiltration model. *Adv. Water Resour.* **2011**, *34*, 257–262. [CrossRef]

21. Barry, D.A.; Parlange, J.Y.; Li, L.; Jeng, D.S.; Crapper, M. Green–Ampt approximations. *Adv. Water Resour.* **2005**, *28*, 1003–1009. [CrossRef]

22. Gerolymatou, E.; Vardoulakis, I.; Hilfer, R. Modeling infiltration by means of a nonlinear fractional diffusion model. *J. Phys. D Appl. Phys.* **2006**, *39*, 4104–4110. [CrossRef]

23. Talbot, C.A.; Ogden, F.L. A method for computing infiltration and redistribution in a discretized moisture content domain. *Water Resour. Res.* **2008**, *44*. [CrossRef]

24. Ogden, F.L.; Lai, W.; Steinke, R.C.; Zhu, J.; Talbot, C.A.; Wilson, J.L. A new general 1-D vadose zone flow solution method. *Water Resour. Res.* **2015**, *51*, 4282–4300. [CrossRef]

25. Yu, H.; Douglas, C.C.; Ogden, F.L. A new application of dynamic data driven system in the Talbot-Ogden model for groundwater infiltration. *Procedia Comput. Sci.* **2012**, *9*, 1073–1080. [CrossRef]

26. Yu, H.; Deng, L.; Douglas, C.C. An extension of the Talbot-Ogden hydrology model to an affine multi-dimensional moisture content domain. *J. Algorithms Comput. Technol.* **2013**, *7*, 271–286. [CrossRef]

27. Simunek, J.; van Genuchten, M.T.; Sejna, M. *The HYDRUS Software Package for Simulating the Two-and Three-Dimensional Movement of Water, Heat, and Multiple Solutes in Variably-Saturated Media, Technical Manual,* 1st ed.; PC Progress s. r. o.: Prague, Czech Republic, 2006.

28. Brooks, R.H.; Corey, A.T. Properties of porous media affecting fluid flow. *J. Irrig. Drain. Div. Proc. Am. Soc. Civ. Eng.* **1966**, *92*, 61–88.

29. Van Genuchten, M.T. A closed-form equation for predicting the hydraulic conductivity of unsaturated soils. *Soil Sci. Soc. Am. J.* **1980**, *44*, 892–898. [CrossRef]

30. Durner, W. Hydraulic conductivity estimation for soils with heterogeneous pore structure. *Water Resour. Res.* **1994**, *30*, 211–223. [CrossRef]

31. Kosugi, K. Lognormal distribution model for unsaturated soil hydraulic properties. *Water Resour. Res.* **1996**, *32*, 2697–2703. [CrossRef]

32. Yong, P.; Huang, J.; Li, Z.; Liao, W.; Qu, L.; Li, Q.; Liu, P. Optimized equivalent staggered-grid FD method for elastic wave modelling based on plane wave solutions. *Geophys. Suppl. Mon. Not. R. Astronom. Soc.* **2016**, *208*, 1157–1172. [CrossRef]

Article

Assessing the Effects of Rainfall Intensity and Hydraulic Conductivity on Riverbank Stability

Toan Thi Duong [1,*], Duc Minh Do [1] and Kazuya Yasuhara [2]

[1] Department of Geotechnics and Infrastructure Development, VNU University of Science, Vietnam National University, Hanoi 100000, Vietnam; ducgeo@gmail.com
[2] Institute for Global Change Adaptation Scienceicas (ICAS), Ibaraki University, 2-1-1 Bunkyo, Mito, Ibaraki 310-8512, Japan; kazuya.yasuhara.0927@vc.ibaraki.ac.jp
* Correspondence: duongtoan109@gmail.com; Tel.: +84-093-454-3261

Received: 15 March 2019; Accepted: 3 April 2019; Published: 10 April 2019

Abstract: Riverbank failure often occurs in the rainy season, with effects from some main processes such as rainfall infiltration, the fluctuation of the river water level and groundwater table, and the deformation of transient seepage. This paper has the objective of clarifying the effects of soil hydraulic conductivity and rainfall intensity on riverbank stability using numerical analysis with the GeoSlope program. The initial saturation condition is first indicated as the main factor affecting riverbank stability. Analyzing high-saturation conditions, the obtained result can be used to build an understanding of the mechanics of riverbank stability and the effect of both the rainfall intensity and soil hydraulic conductivity. Firstly, the rainfall intensity is lower than the soil hydraulic conductivity; the factor of safety (FOS) reduces with changes in the groundwater table, which is a result of rainwater infiltration and unsteady state flow through the unsaturated soil. Secondly, the rainfall intensity is slightly higher than the soil hydraulic conductivity, the groundwater table rises slowly, and the FOS decreases with both changes in the wetting front and groundwater table. Thirdly, the rainfall intensity is much higher than the soil hydraulic conductivity, and the FOS decreases dominantly by the wetting front and pond loading area. Finally, in cases with no pond, the FOS reduces when the rainfall intensity is lower than hydraulic conductivity. With low hydraulic conductivity, the wetting front is on a shallow surface and descends very slowly. The decreasing of FOS is only due to transient seepage changes of the unsaturated soil properties by losing soil suction and shear strength. These obtained results not only build a clearer understanding of the filtration mechanics but also provide a helpful reference for riverbank protection.

Keywords: riverbank stability; rainfall intensity; hydraulic conductivity

---

1. Introduction

Rainfall is one of the main factors causing slope failure in tropical areas. The effects of rainfall properties on slopes have been studied in large amounts of research that have analyzed slope stability by both investigation and simulated models. By the simulation of different conditions of slopes and rainfall boundaries, previous research has indicated the changes of soil unsaturated properties, such as the reduction of suction, shear strength, and the increasing of hydraulic conductivity and pore-water pressure [1–20]. Finally, these changes cause slope failure. Before rain, the slope area above the groundwater table is considered as being in a partly unsaturated and dry state near the surface [1–4]. The unsaturated area reduces by rainfall water infiltration during and after the rainfall event. The change of the area from unsaturated to saturated is caused by the advancement of the wetting front from the surface [4,5,9,13,21–23] and groundwater table from depth [6–8,10–20]. Those processes were found to be the primary factors controlling the instability of slopes due to rainfall and were greatly affected by rainfall intensity (RI) and soil properties, especially by unsaturated soil hydraulic conductivity [4,17–36]. The unsaturated soil hydraulic conductivity (HC) controls the transient seepage, the depth of rainfall infiltration, the changes in pore pressure during the rainfall event, and finally, affects the FOS. The effects of soil hydraulic conductivity on slope stability in the rainy season are usually assessed from the point of view of three topics: (1) the effects of changes in the value of soil hydraulic conductivity and rainfall intensity [17–30]; (2) the soil anisotropic HC and hydraulic hysteresis [31–35]; (3) the failure delay phenomenon due to the defense of HC and RI [36]. This research focuses on the overview and analysis of the first of these topics.

The effects of the HC on the FOS were specifically simulated by [25,26]. Those research works reported that high hydraulic conductivity led to rapid saturation. Additionally, infiltration causes the wetting front to quickly shift downward. This shift causes water to contact the underlying impermeable soil, leading to a rapid rise in pore-water pressure and the formation of a perched water table. The slope reaches full saturation and experiences a reduction in soil resistance. Consequently, the factor of safety rapidly decreases [26]. High-intensity rainfall has large effects on the slope if the soil slope has high hydraulic conductivity ($Ks \geq 10^{-4}$ m/s) and when the slope has poorly drained soils (i.e., $Ks \leq 10^{-6}$ m/s) [25].

The simulated analyses were also carried out with different boundaries of rainfall. When building the relationship of rainfall intensity and unsaturated hydraulic conductivity by the ratio of RI/HC, the previous research analyzed the RI = HC [5], RI $\geq$ HC [19,22]; and RI < HC [8,17,21,23,26,29] cases. Setting up a boundary with RI = HC or RI $\geq$ HC and no pond [5,19,22], the development of a wetting front from the crest of the slope and the reduction of soil suction during rainfall were found to dominantly affect the slope stability. The development of the defense of the wetting front under increased rainfall intensity shortens the time required for the wetting front to reach the pore-water pressure and moisture content sensors. The growing of groundwater has not been mentioned much. The same trend was also found in [17,21,23], which had the boundary of RI < HC. The larger the coefficient of permeability is, the greater the depth of the wetting front is. The opposite mechanism is seen in [8,29] with a boundary of small RI and high initial saturation and hydraulic conductivity, and the pore-water pressure increases gradually from the deep part to the crest of the slope [8,29]. For a slope with a larger HC, the slope failures possibly take place under rainfall with a shorter duration and a greater intensity [8]. Those research works also concluded that when the rainfall intensity is greater than the saturated hydraulic conductivity of the surface soils, a runoff occurs along the slope surface [8,22].

Although those previous research works had unobvious objectives focusing on building the effects of hydraulic conductivity, the results clearly showed that the slope stability was significantly affected by the soil hydraulic conductivity and the rainfall intensity boundary. The ratios of RI/HC were divided into three cases; however, each previous research work only concentrated on one of those cases. Moreover, there was a difference regarding the development of the pore-water pressure by the rising groundwater table [8,29] or by a wetting front from the surface [21,23] when those analyses had the same initial conditions. Further research should cover all three cases of different RI/HC ratios.

A riverbank is a special slope that relates closely to hydraulic dynamics, not only from rainfall infiltration from the surface concerning the space above the slope, but also significantly from the fluctuation of the river water level and groundwater table [27,37–43]. Past research works assessed the effects of rainfall on riverbank stability using the indirect process of transient seepage due to water level changes and assumed that rainfall has no effect on the riverbank surface.

By reviewing previous papers, it was obvious that the hydraulic conductivity has great effects on slope stability during rainfall events. However, some limitations were found in the reviewed research. The effects of the different RI/HC values were not specific to a researcher, and there was a difference in the previous discussion on seepage mechanics, such as the changes in the wetting front and the groundwater table. Moreover, research on the effects of hydraulic conductivity on riverbank slope stability was performed in only a few papers. Therefore, the objective of this paper is to build the mechanics of riverbank failure with different rainfall infiltration and soil hydraulic conductivity models. The numerical analyses by the GeoSlope program with both SEEP/W and SLOPE/W moduli are applied in this study.

2. Materials and Methods

The case study is a riverbank area in the Red River in Hanoi city, Vietnam. The Red River originates in Yun-nan, China, and flows through northwest Vietnam. The total length of the Red River in Hanoi is about 90 km with about 40 km of riverbank inside the urban area of Hanoi, where the population density and density of houses built near the natural riverbank are highest. These areas are located in high river terraces, outside the river dike, and near natural banks. In the annual rainy season, the destruction of local houses, roads, and land-use along the Red River bank has occurred.

The selected riverbanks are almost natural banks, and some are supported in the toe by vegetation. The riverbanks in these locations have either collapsed or have high failure potential due to fluctuations in the river water level, river water flow stress, rainfall, and human activities. Figure 1 shows the road near the riverbank (left), and the natural riverbank (right), which was broken in the rainy season of 2016 and 2017.

Figure 1. Some current problems near the riverbank in the study area.

2.1. Field Investigations and Soil Properties

The field investigation was performed during both the dry and the rainy seasons to describe the status of the riverbank and the river water level changes. Field data measurement and collection included bank geometry (i.e., height, slope), alluvial area, current river water level, and soil samples. The monitoring data, which included groundwater level, river water level fluctuation, and rainfall, were also collected from the National Meteorology Station.

The soil properties included soil physical properties such as water content, density, and grain size. The saturated hydraulic conductivity and shear strength were obtained in the geotechnical laboratory of VNU University of Science, Vietnam National University, Hanoi. The unsaturated soil properties, such as soil suction, were determined by using a pressure plate apparatus in the geotechnical laboratory of Ibaraki University, Japan, as shown in [42]. Based on the soil samples collected along the riverbank in the Hanoi area and the soil properties experiment, the riverbank in this area is quite homogeneous, with a silt or silty-clay layer in the bank layer and fine sand from the toe of the riverbank to the sediments. The silt layer is composed of less than 20% fine sand, 30–70% silt, and 10–30% clay. This paper selects one section of the soil bank to analyze riverbank stability, as shown in Table 1. Figures 2 and 3 present the suction and unsaturated hydraulic conductivity, respectively. The unsaturated shear strength and hydraulic conductivity were estimated by using the models of Vanapalli (1996) [44] and Van Genuchten (1980) [45] in the GeoSlope program [46,47].

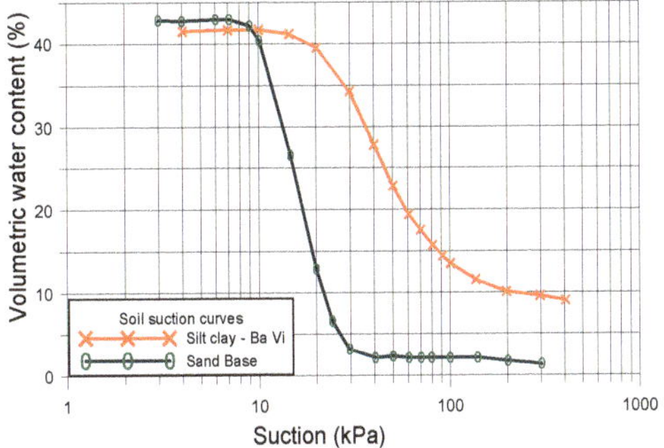

Figure 2. Soil–water characteristic curves.

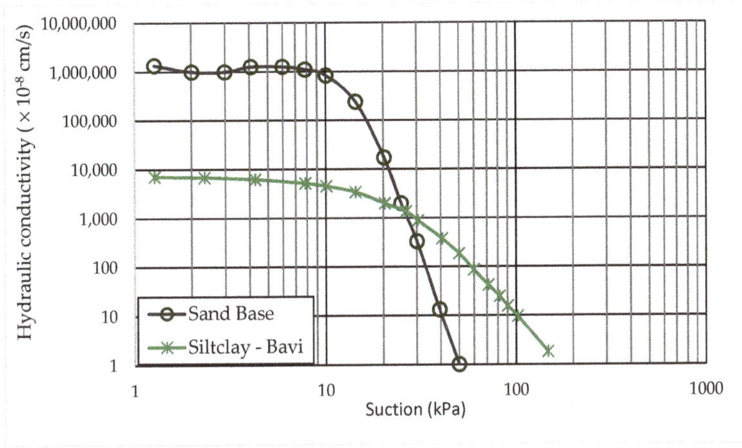

Figure 3. The unsaturated hydraulic conductivity curves.

Table 1. Soil properties used in riverbank stability analysis in Ba Vi area.

Riverbank Soil Properties		Silt-Riverbank Soil	Fine Sand Base
Depth of layer		10–1.5m	From 1.5 m to depth
Grain size (%)	Coarse sand: 1–0.25 mm		18.38
	Fine sand: 0.25–0.075 mm	13.46	80.91
	Silt: 0.075–0.005 mm	71.03	0.71
	Clay < 0.005 mm	15.51	0
Natural water content (%)		18.5	4.72
Dry density (kN/m^3)		15.0	
Specific gravity		2.62	2.68
Liquid limit (%)		34.5	
Plastic limit (%)		21	
Liquid index		1.23	
Soil classification		ML	SP
Saturated volume water content		42	29
Air-entry value (kPa)		20.05	9.3
Residual suction (kPa)		90	25
Residual volumetric water content (%)		10	4

Table 1. *Cont.*

Riverbank Soil Properties	Silt-Riverbank Soil	Fine Sand Base
a	28.39	10.33
n	4.205	18.89
m	0.72	0.53
Max slope	1.33	2.58
Hydraulic conductivity (cm/s)	7.39×10^{-5}	1.69×10^{-2}
Cohesion force ((kPa)	5.0	0
Internal friction angle (o)	32	30

2.2. Numerical Model Framework

This paper uses the commercial GeoSlope program (GeoSlope International Ltd.) [46–49] as a numerical model to analyze riverbank stability. GeoSlope is one of the most useful and widely used programs in slope and riverbank stability analysis in many kinds of research [1–5,12,14,15,24–26]. The present paper uses a pair of analyses of transient seepage in SEEP/W and slope stability in SLOPE/W in the GeoSlope program.

In general, by using the SEEP/W, the mechanism describes specifically the transient seepage by rainfall infiltration. The obtained results from SEEP/W, which include pore-water pressure distribution and changes in soil properties, become the input data for the next modulus SLOPE/W for analyzing the riverbank stability. The result of the FOS indicates the riverbank stability when FOS is higher than one. The results and discussion focus on building the relationships between the FOS and the different conditions of initial saturation, rainfall intensity, and soil hydraulic conductivity.

The field investigation, laboratory testing, and monitoring of the support input data included three factor groups: the riverbank geometry, the soil properties, and the hydraulic conditions (such as the river water level and rainfall intensity). The riverbank stability analysis was performed for different initial saturation conditions, soil hydraulic conductivity, and rainfall intensity. The initial saturation conditions were set up by the initial river water level and the capillary height or the maximum negative head. In SEEP/W, the maximum negative head can be used to build the assumption of the predetermined negative pore pressure profile, and then the saturation condition can be set up. Other simulated factors such as the soil hydraulic conductivity and rainfall intensities were installed as the difference between the soil properties and boundaries. The conditions of the riverbank, hydraulics, and boundaries for different scenarios of the initial saturation and rainfall intensity are described in more detail below.

2.2.1. Riverbank Geometry and Hydraulic Boundary Condition

This paper uses a riverbank configuration with the initial conditions shown in Figure 4. The riverbank has a slope angle of 52 degrees and a height of 10 m. The riverbank is homogeneous silty soil with a sand layer underneath. The analyzed soil properties are shown in Table 1 and Figures 2 and 3.

The boundary conditions of river water level (RWL) and rainfall intensity were set up in the SEEP/W model. Using the riverbank configuration shown in Figure 4, the initial RWL was the boundary in the river site from the bottom to 3 m, and the function RWL–time was the boundary of the entire lateral riverbank. The rainfall was the boundary on the surface of the riverbank. Both scenarios with and without the function "potential seepage face review" were used with the rainfall boundary. Without the function "potential seepage face review", the rainwater would infiltrate into the soil as long as the rainwater exceeded the infiltration capacity; under these conditions, a pond would present above the riverbank surface when there was excess rainwater. This condition often occurs at some riverbank sites that have a pond or low areas near the riverbank and during floods. If the "potential seepage face review" was included, the pond would not exist, and the excess rainwater would run off.

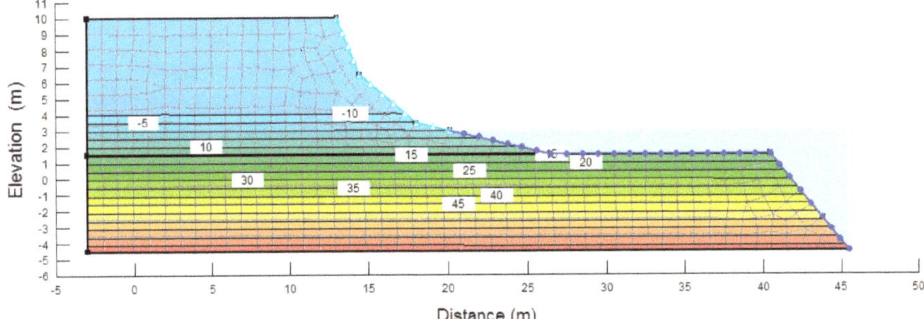

Figure 4. The initial riverbank configuration.

The rainfall intensity was determined based on daily monitoring data. The rainy season is from mid-June to mid-September, and high rainfall often occurs in August. Figure 5 shows the variation in the rainfall intensities (RI, mm/h) of several rainy days in the rainy month (August 2016), and Figure 6 shows the changes in daily rainfall and RWL monitored during the rainy season from 1 July to 31 August where some sites in the Red River bank broke. It can be seen that the rainfall intensity ranged from 0–50 mm/h. To simulate the effect of rainfall intensity on riverbank stability, three rainfall intensities were used: RI = 10 mm/h; RI = 30 mm/h; and RI = 50 mm/h. Based on the RWL data, the initial RWL was set as 3 m.

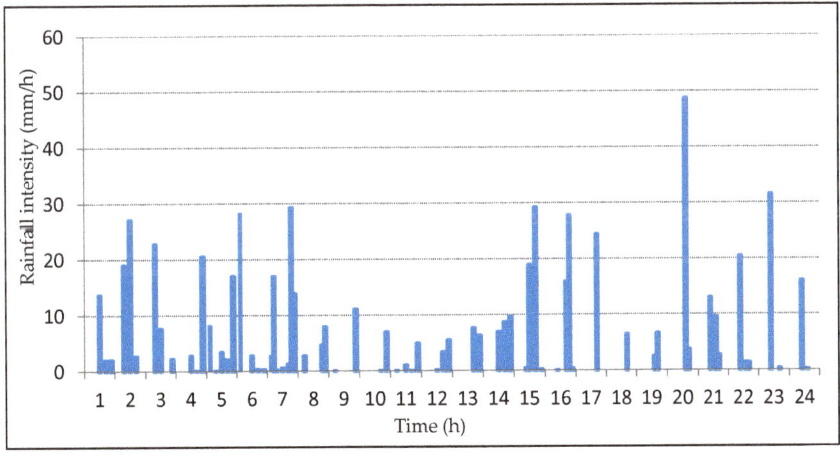

Figure 5. The hourly rainfall with high rainfall intensity for some days in August 2016.

The initial pore-water pressure and saturation conditions were set up by using the function of the maximum negative pressure head in SEEP/W. This function was built with knowledge of the pore pressure in unsaturated soil. Soil regions above the groundwater table were divided into two sub-regions: a dry zone near the surface and a partly saturated zone near the groundwater table. The pore-water pressure graph was linear and negatively sloped from the groundwater table to the maximum negative head. This meant that the negative pore-water pressure near the surface may have become too high in the dry zone. In fact, the soil was never completely dry and always retained some amount of water. The small surface flux had the effect of changing the pore-water pressure profile. Figure 7 shows a pore-water pressure profile with a non-dry surface condition in which the negative pore-water pressure was negatively linearly sloped to a maximum negative pore-water pressure and remained constant at a value in response to soil water content. The magnitude of the maximum negative pore-water pressure was dependent on the shape of the hydraulic conductivity function, and to a lesser extent on the rate of infiltration. In SEEP/W, setting the maximum negative head can indicate the field pore pressure profile. Based on the investigation and experimental data of the soil water content and soil suction curve, the value pore pressure or saturation degree could be determined.

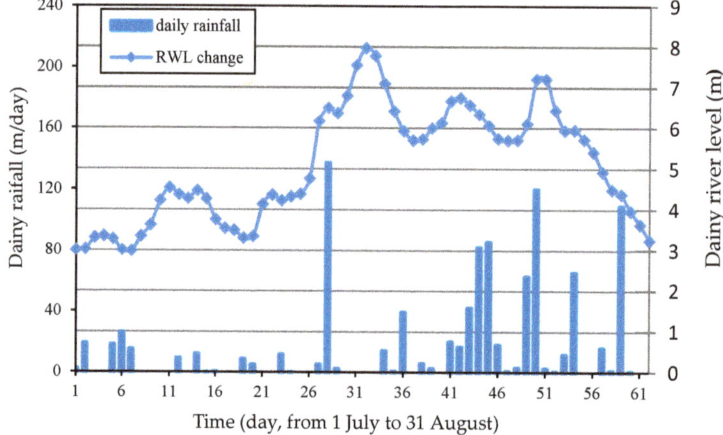

Figure 6. Daily river water level (RWL) and daily rainfall from 1 July to 31 August 2016.

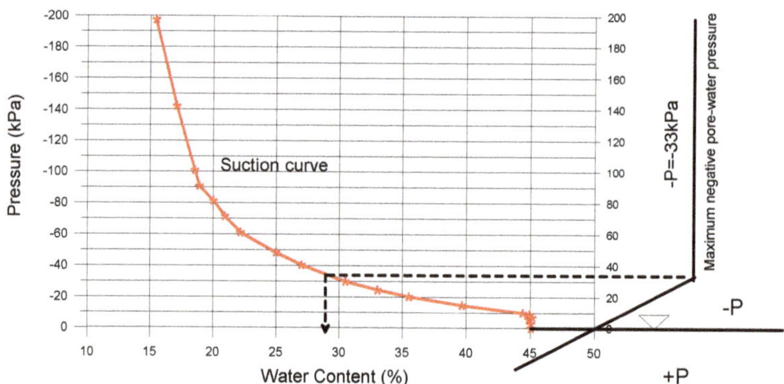

Figure 7. Calculation pore-pressure condition by saturation condition.

To simulate the effect of the initial pore pressure and saturation conditions, we set the different initial negative pore pressures to 15 kPa and 33 kPa, respectively. At those pressures, the relative soil water contents were 41% and 33%, and the saturation degrees were 87% and 70% respectively. Those were average experimental values in the beginning of the rainy season and in the rainy season.

The initial riverbank soil had a saturation hydraulic conductivity of $Ks = 7.39 \times 10^{-5}$ cm/s (Table 1). To simulate the effects of soil hydraulic conductivity on the processes of rainfall infiltration and riverbank stability, three values of saturation hydraulic conductivity were used to represent the initial saturated hydraulic conductivity: $Ks = 7.39 \times 10^{-3}$ cm/s, $Ks = 7.39 \times 10^{-4}$ cm/s, and $Ks = 7.39 \times 10^{-5}$ cm/s. A hydraulic conductivity of $Ks = 7.39 \times 10^{-3}$ cm/s was considered to show high conductivity (cases H_i); $Ks = 7.39 \times 10^{-4}$ cm/s was considered medium hydraulic conductivity (cases M_i); and $Ks = 7.39 \times 10^{-5}$ cm/s was considered low hydraulic conductivity (cases L_i). Figure 8 shows three unsaturated hydraulic conductivity curves that correspond to the three hydraulic conductivity values listed above and to the same suction properties for silty soil in the Red River bank.

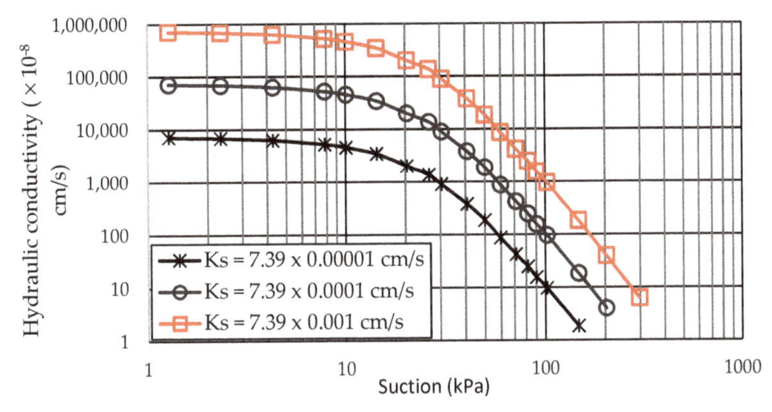

Figure 8. The different unsaturated hydraulic curves used in analyses.

2.2.2. Cases Used in Analysis

Table 2 describes the numbers of cases we analyzed to simulate the riverbank stability in the different initial suction conditions or negative pressure, the rainfall intensity, and the hydraulic conductivity, in which cases H-1-1, H-1-2, H-1-3 were the names of case studies with high saturation hydraulic conductivity (Ks = 7.39 $\times$ 10^{-3} cm/s) at the initial saturation degree of 70% at three rainfall intensities of 10 mm/h, 30 mm/h, and 50 mm/h, respectively. Using this labeling for cases, there were 18 analyzed cases, as shown in Table 2.

Table 2. Cases used to analyze Ba Vi riverbank.

Saturated Degree (%)	Vol. Water Content (%)	Soil Suction (kPa) in Ba Vi	The Unsaturated Hydraulic Conductivity (HC, cm/s) Ba Vi	Name of Cases Versus HC	Name of Cases			The Ratio RI/HC at		
					10 mm/h	30 mm/h	50 mm/h	10 mm/h	30 mm/h	50 mm/h
70	33	33	7×10^{-4}	H-1	H-1-1	H-1-2	H-1-3	0.38	1.19	1.85
87	41	15	3×10^{-3}	H-2	H-2-1	H-2-2	H-2-3	0.09	0.28	0.46
70	33	33	7×10^{-5}	M-1	M-1-1	M-1-2	M-1-3	3.85	11.86	18.57
87	41	15	3×10^{-4}	M-2	M-2-1	M-2-2	M-2-3	0.93	2.78	4.63
70	33	33	7×10^{-6}	L-1	L-1-1	L-1-2	L-1-3			
87	41	15	3×10^{-5}	L-2	L-2-1	L-2-2	L-2-3	9.26	27.78	46.30

Where rainfall intensities (RI) = 10 mm/h = 2.7 $\times$ 10^{-4} cm/s; RI = 30 mm/h = 8.3 $\times$ 10^{-4} cm/s; RI = 50 mm/h = 1.3 $\times$ 10^{-3} cm/s.

3. Results

3.1. Effects of the Initial Saturation Condition

Figure 9 shows the results of FOS versus time with the saturation degrees of 70% and 87% and RI = 10 mm/h. These were cases H-1-1, H-2-1, M-1-1, and M-2-1. In these cases, the rainfall intensity was lower than the hydraulic conductivity when the saturation degree was 87%, and the rainfall intensity was slightly higher than the hydraulic conductivity when the saturation degree was 70%. Most of the rainfall water infiltrated into the soil and caused the increase of pore-water pressure by the raising of the groundwater table. The FOS results indicated the effect of the saturation condition and hydraulic conduction on riverbank stability. The higher saturation condition and higher hydraulic conductivity caused higher pore pressure; then, FOS decreased more quickly and obtained a lower value (Figure 9).

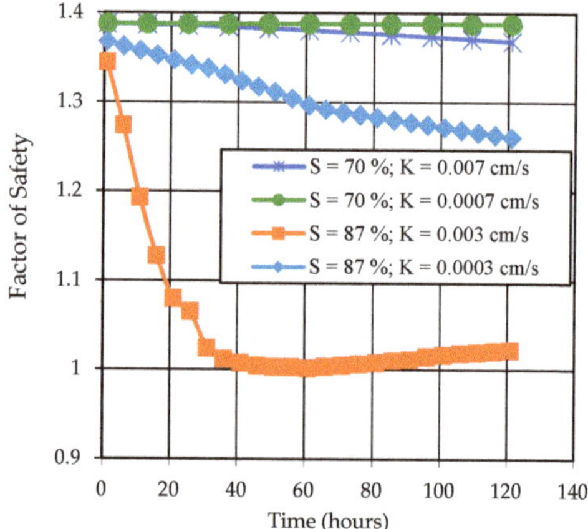

Figure 9. The factor of safety (FOS) results at an RI = 10 mm/h and with a different saturation degree and hydraulic conductivity.

When S = 87%, the FOS had obviously different trends that depended on the rainfall intensity and soil hydraulic conductivity. However, it was an insignificant change in the case of the analysis where S = 70%. Therefore, the below results and discussion mention only the case where S = 87 %.

3.2. Effects of Rainfall Intensity and Hydraulic Conductivity on Riverbank Stability

Firstly, both the rainfall intensity and rainfall accumulation affected the FOS. With higher rainfall intensity and rainfall accumulation, the FOS decreased to a lower value (Figure 10). That result was also found in most of the previous research that mentions rainfall intensity [21–30]. Figure 10A1,A2,B1,B2,C1,C2 shows the changes of FOS not only by rainfall intensity and rainfall accumulation but also by soil hydraulic conductivity in cases with a pond and no pond on the surface. In Figure 10, the red dashed line where FOS was equal to one on the graphs meant that riverbank failure occurred. The effects of hydraulic conductivity on the FOS were as follows:

When Ks = 7.39 × 10^{-3} cm/s, the change in FOS had the same trend in both boundary cases. In three cases of different RIs, the FOS decreased at the same rate from the beginning of the rainfall event to approximately 30 h. After 30 h of rain, the FOS varied with the different rainfall intensities. The higher the RI was, the lower the FOS was. The riverbank was stable when RI = 10 mm/h; however, riverbank failure occurred after 36 h with RI = 50 mm/h, and after 40 h with RI = 30 mm/h. The results had the same trend as those obtained by Rahimi et al [25]. There, the FOS also decreased rapidly in the beginning of the rain event, and then FOS insignificantly changed after a threshold of RI. Once the RI reached the threshold RI, the FOS did not change any more due to an excess of rainwater run off [25].

For Ks = 7.39 × 10^{-4} cm/s, the range of FOS in cases with a pond was from 1.36 to 0.39 (Figure 10B1), and from 1.39 to 1.03 in cases with no pond (Figure 10B2). In both cases, the FOS change was insignificant with an RI of 10 mm/h; however, the FOS decreased rapidly after 55 h and 110 h where RI = 50 mm/h and RI = 30 mm/h, respectively. In the case with a pond, the riverbank failure occurred at 60 h and 115 h with RI = 50 mm/h and RI = 30 mm/h, respectively. In the case with no pond, riverbank failure only occurred when RI = 50 mm/h.

For Ks = 7.39 × 10^{-5} cm/s, the FOS obviously decreased in the rainfall boundary with a pond (Figure 10C1), but that changed only slightly in the case of a boundary with no pond (Figure 10C2). In the case with a pond, the FOS decreased as the rainfall intensity and accumulation increased from 1.38 to 0.2. At higher rainfall intensity, riverbank failure occurred quickly. Riverbank failure occurred after 16 h and 26 h for rainfall intensities of 50 mm/h and 30 mm/h, respectively. In the case of no pond, the FOS decreased indistinguishably in small ranges with all rainfall intensities from 1.37 to 1.35, and the riverbank was stable after five days of rain.

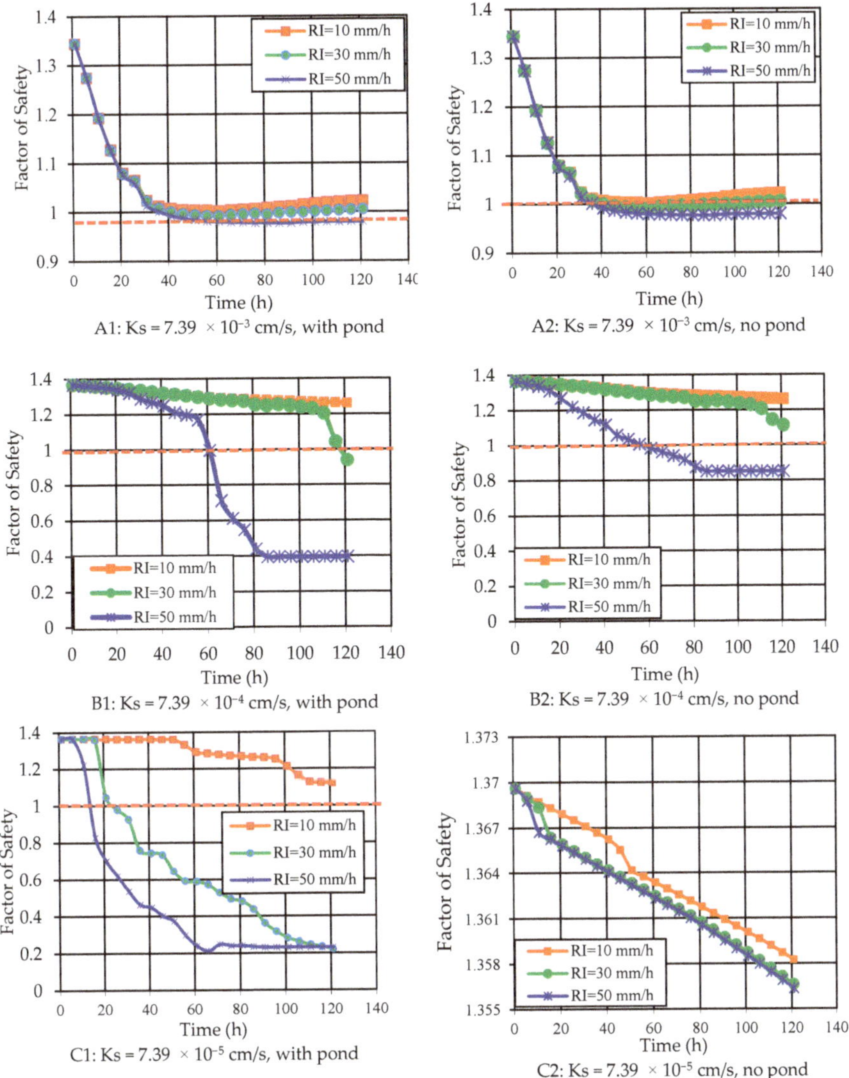

Figure 10. The change of FOS with different rainfall intensities and hydraulic conductivity with a pond and with no pond.

These results indicated that the drainage condition (pond or no pond) greatly affected the change of FOS. In the case of no pond, the present results regarding the change of FOS with different RIs and HC were the same as the results obtained in the same condition with no pond in [21–23,27–30]. The potential of riverbank failure was higher when the riverbank had higher hydraulic conductivity. On the contrary, the potential of riverbank failure with low hydraulic conductivity was higher than that with high hydraulic conductivity when a pond was present on the surface. With a pond on the surface, the excess rainwater not only built a loading pressure but also created a wetting front;

then, the FOS decreased more quickly. With HC < RI, the FOS had a lower value in soils with lower hydraulic conductivity (Ks = 3 × 10^{-5} cm/s) than in soils with higher hydraulic conductivity (Ks = 3 × 10^{-4} cm/s) because the excess rainwater in the first case was much more than in the latter case. In general, it is rare to have a pond on the surface of a riverbank or slope. However, when the riverbank slope has a high saturation degree and the soil suction is completely lost, any charge-loading in the surface will cause an unbalanced loading and riverbank slope failure to occur.

3.3. Effects of Rainfall Intensity and Hydraulic Conductivity on the Water Infiltration Mechanism

The mechanics of rainfall infiltration for different cases of RI and HC are indicated here by the changes of pore-water pressure and the wetting line. Pore-water pressure increased with the raising of the groundwater table, and the wetting front was linearly propagated from the surface and separated the saturated area and the initial unsaturated area by rainwater infiltration.

In the case with a pond on the slope surface, the rainfall infiltration occurred in both processes: (1) the pore-water pressure or the groundwater table rose, and (2) the wetting front propagated. The changes in the groundwater table or the wetting front depended on the rainfall intensity and the soil hydraulic conductivity.

The rainfall infiltration caused the raising of the groundwater table when the hydraulic conductivity was higher than the rainfall intensity. Figure 11 shows the groundwater table versus raining time for the case H-1-1, which had an RI/HC = 0.09. The rainwater infiltrated and moved quickly into the soil and could raise the groundwater table. The transient seepage area above the groundwater may not even have obtained full saturation, as shown in [17]. The cases of H-1-1, H-2-2, H-2-3, and M-1-1 with the ratios of RI/HC = 0.38; 0.28, 0.46, and 0.93, respectively, had the same mechanics. In these cases, an increase of pore-water pressure was the main reason causing the decrease of the FOS and riverbank failure. These results had the same trend as shown in [8,29] when the soil hydraulic conductivity was higher than the rainfall intensity.

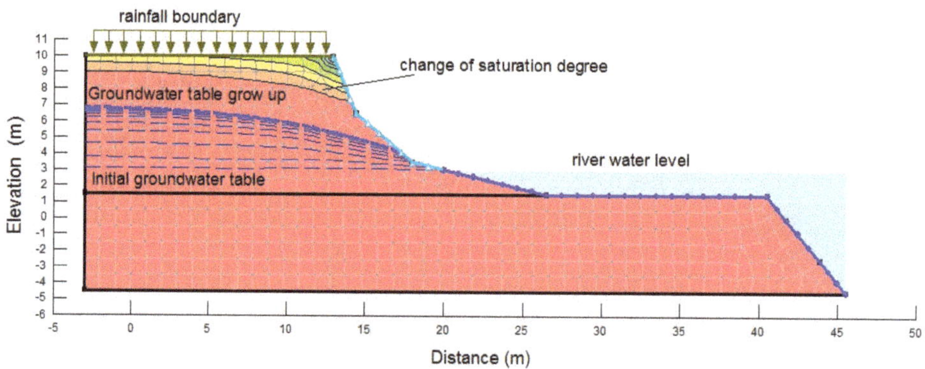

Figure 11. The raising of the groundwater table when the RI/HC is less than one.

The second process, in which the rainfall infiltration caused the wetting front to descend, occurred when the hydraulic conductivity was lower than the rainfall intensity or the ratio of RI/HC was higher than one (i.e., the cases of M-2-2, M-2-3, L-2-1, L-2-2, and L-2-3). With a high ratio of RI/HC or low hydraulic conductivity, as in cases L-2-1, L-2-2, and L-2-3, the rainwater infiltrated very slowly into the riverbank soil and did not cause a change in groundwater. Because of the slow transient seepage and the high rainfall intensity, an amount of excess water created a pond on the surface. When a pond was present, the wetting line appeared and descended deeper into the riverbank soil as rainfall accumulation increased. Figure 12 shows the wetting front descending in the case of low hydraulic conductivity, i.e., Ks = 7.39 × 10^{-5} cm/s, with the pond above the riverbank surface when the rainfall flux exceeded the soil infiltration capacity. The increased wetting depth caused the fully saturated zone to expand to nearly the riverbank surface. The rainfall intensity greatly affected the depth of the wetting front and the height of the pond. In these cases, the rainfall intensity and changes in the saturation condition caused by the wetting front were the main factors affecting the FOS and the riverbank stability.

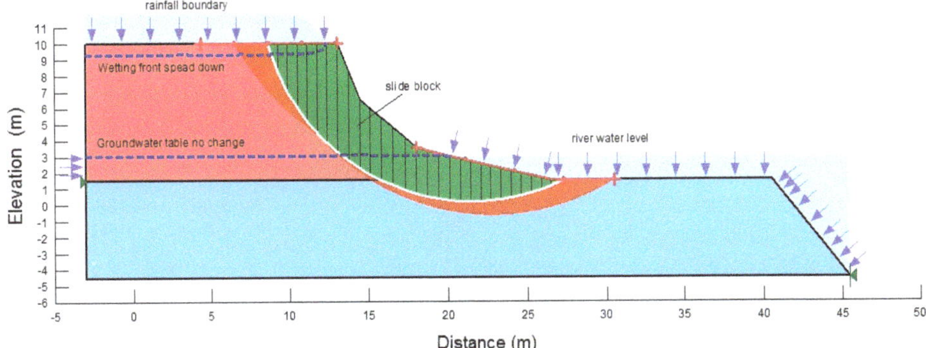

Figure 12. The wetting front descending when the RI is much higher than HC.

When the rainfall intensity was slightly higher than hydraulic conductivity (as in cases M-2-2 and M-2-3), the riverbank became saturated from both the groundwater table and the wetting front, as shown in Figure 13. Similar to the case of high hydraulic conductivity, the groundwater table rose by transient seepage through the unsaturated area. However, the groundwater rose at a slow rate. The wetting front appeared when excess rainwater was present on the surface. With only the rising of the groundwater table, the FOS decreased slowly, but the FOS changed more quickly when the wetting front descended. In this case, both the groundwater table and wetting front influenced the FOS and the riverbank stability.

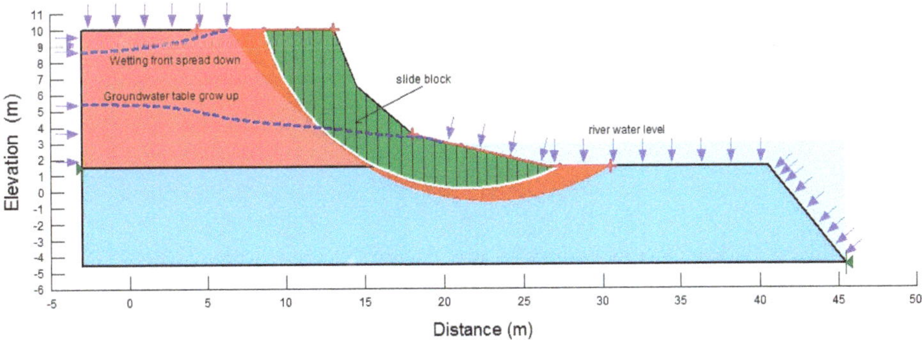

Figure 13. The rising of the groundwater table when the RI is lightly higher than HC.

In cases where the riverbank had good drainage conditions and no pond on the surface, the excess rainwater ran off the slope. When RI < HC (cases H-1-1, H-2-1, H-2-1, H-2-3, and M-2-1), the change of groundwater was the same as in the cases with a pond on the surface.

In cases with low hydraulic conductivity, (cases M-2-2, M-2-3, L-2-1, L-2-2, and L-2-3), the rainwater seepage occurred very slowly. The excess water ran off the slope, and the wetting front also spread slowly. In a short time, the groundwater table and the wetting front did not significantly increase. Then, the FOS slightly decreased (Figure 10C2). The decreasing of the FOS was due to the transient seepage change of the unsaturated soil properties as well as losing soil suction. This result also matched that in [2,5,19,22] when the rainfall intensity was smaller or equal to the soil hydraulic conductivity.

3.4. Effects of Rainfall Infiltration and River Water Level Fluctuation

The effects of multiple rainfall events and RWL in the rainy season are shown in Figure 14. The change in FOS was nearly the same as the change in RWL, which meant that the FOS increased with the increase in RWL, and the FOS decreased with the decrease in RWL. These results agreed with those of previous studies [39–42], which studied the effect of RWL on FOS. When the riverbank had a soil hydraulic conductivity lower than 10^{-4} cm/s, the FOS always depended on the change in RWL because of confining pressure [42]. The effect of rainfall on the riverbank was insignificant compared with the effects of changes in RWL. The results from the long-term analysis during the rainy season showed that low-intensity rainfall (less than 140 mm/day, average 5.8 mm/h) caused the FOS to decline at a low rate. Riverbank failure occurred only when the riverbank had a high slope and had cracks in the riverbank surface. In fact, there was a high slope angle and some cracks along the riverbank in the study area. In general, a riverbank with a high slope angle and some cracks has a high potential of failure during the rainy season. Moreover, the change in RWL was the factor with the greatest influence on the FOS of the riverbank. The riverbank was more damaged when multiple factors occurred together.

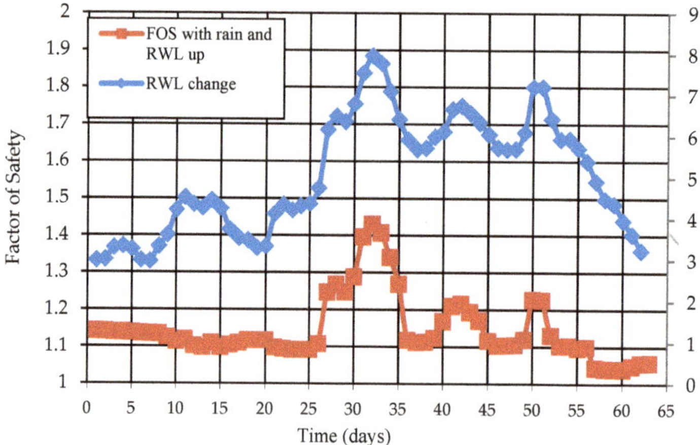

Figure 14. The FOS change with fluctuations of RWL and rainfall in the long-term.

4. Conclusions

The initial conditions such as the saturation degree and the groundwater table are the first important factors for riverbank slope stability in the rainy season. Setting the initial conditions leads to different mechanisms of infiltration, seepage, and changes of groundwater pressure.

In highly saturated conditions, the FOS decreases with the increase in rainfall intensity and accumulation. During a rainfall event, the rainwater infiltrates and affects the riverbank stability through two processes—changes in pore-water pressure and a wetting front controlled by the rainfall intensity and the hydraulic conductivity.

When the rainfall intensity is lower than soil hydraulic conductivity, the rainwater infiltrates, and transient seepage occurs through the unsaturated area, causing the groundwater table to rise. In soil with a higher hydraulic conductivity and rainfall intensity, the groundwater table rises quickly, leading to a higher potential of riverbank failure. Riverbank failure often occurs under high-intensity rainfall.

When the rainfall intensity is slightly higher than the soil hydraulic conductivity, the groundwater table rises slowly, and the FOS decreases slowly in the early period after raining. With a pond on the surface, the wetting front appears when there is excess rainfall water. The FOS decreases more quickly by the development of both the wetting front and the groundwater table convergence.

When the rainfall intensity is much higher than the soil hydraulic conductivity, the groundwater does not change in a short time. The wetting front descends slowly, and the height of the pond increases quickly with higher rainfall intensity. The wetting front and pond loading area are the main factors causing FOS changes.

In cases with no pond, the wetting front is on a shallow surface and descends very slowly, and the rainfall water transient seepage occurs slowly. In a short time, the groundwater table and the wetting front do not significantly increase; thus, the FOS slightly decreases. The decreasing of FOS is due to transient seepage changes of unsaturated soil properties as well as losing soil suction.

During long-term events with low-intensity rainfall, i.e., less than 10 mm/h, the FOS primarily depends on the river water level. A riverbank with a high slope angle and cracks with high hydraulic conductivity will have a higher potential of riverbank failure.

Author Contributions: Conceptualization, T.T.D. and D.M.D.; Methodology, T.T.D.; Software, T.T.D.; Validation, T.T.D., D.M.D. and K.Y.; Formal Analysis, T.T.D.; Investigation, T.T.D., D.M.D.; Resources, T.T.D.; Data Curation, T.T.D.; Writing-Original Draft Preparation, T.T.D.; Writing-Review & Editing, T.T.D.; Visualization, T.T.D.; Supervision, K.Y.; Project Administration, T.T.D.; Funding Acquisition, T.T.D.

Funding: This research was funded by the project Code 105.08-2015.24, which was sponsored by Nafosted, Ministry of Science and Technology, Vietnam.

Acknowledgments: This paper was completed with the support of the project Code 105.08-2015.24, which was sponsored by Nafosted, Ministry of Science and Technology, Vietnam, and the support by Geotechnical Laboratory, Ibaraki University, Japan to determine unsaturated soil properties. The authors express our sincere gratitude for these supports.

Conflicts of Interest: The authors declare no conflict of interest.

References

1. Gasmo, J.; Rahardjo, H.; Leong, E.C. Infiltration effects on stability of a residual soil slope. *Comput. Geotech.* **2000**, *26*, 145–165. [CrossRef]
2. Iverson, R.M. Landslide triggering by rain infiltration. *Water Resour. Res.* **2000**, *36*, 1897–1910. [CrossRef]
3. Rahardjo, H.; Leong, E.C.; Rezaur, R.B. Studies of rainfall-induced slope failures. In Proceedings of the National Seminar, Slope 2002, Bandung, Indonesia, 27 April 2002; pp. 15–29.
4. Rahardjo, H.; Ong, T.H.; Rezaur, R.B.; Leong, E.C. Factors controlling instability of homogeneous soil slopes under rainfall. *J. Geotech. Geoenviron. Eng.* **2007**, *133*, 1532–1543. [CrossRef]
5. Kim, J.; Jeong, S.; Park, S.; Sharma, J. Influence of rainfall-induced wetting on the stability of slopes in weathered soils. *Eng. Geol.* **2004**, *75*, 251–262. [CrossRef]
6. Lu, N.; Godt, J. *Hillslope Hydrology and Stability*; Cambridge University Press: New York, NY, USA, 2013.
7. Gofar, N.; Lee, M.L.; Asof, M. Transient seepage and slope stability analysis for rainfall-induced landslide: A case study. *Malays. J. Civ. Eng.* **2006**, *18*, 1–13.
8. Cai, F.; Ugai, K. Numerical analysis of rainfall effects on slope stability. *Int. J. Geomech.* **2004**, *4*, 69–78. [CrossRef]
9. Huang, C.C.; Lo, C.L.; Jang, J.S.; Hwu, L.K. Internal soil moisture response to rainfall-induced slope failures and debris discharge. *Eng. Geol.* **2008**, *101*, 134–145. [CrossRef]
10. Egeli, I.; Pulat, H.F. Mechanism and modelling of shallow soil slope stability during high intensity and short duration rainfall. *Sci. Iran.* **2011**, *18*, 1179–1187. [CrossRef]
11. Suryo, E.A. Real-Time Prediction of Rainfall Induced Instability of Residual Soil Slopes Associated with Deep Cracks. Doctoral Thesis, School of Earth, Environment and Biological Science Science and Engineering Faculty Queensland University of Technology February, Brisbane, Australia, 2013.
12. Yunusa, G.H.; Kassim, A.; Gofar, N. Effect of surface flux boundary conditions on transient suction distribution in homogeneous slope. *Indian J. Sci. Technol.* **2014**, *7*, 2064–2075.
13. Oh, S.; Lu, N. Slope stability analysis under unsaturated conditions: Case studies of rainfall-induced failure of cut slopes. *Eng. Geol.* **2015**, *184*, 96–103. [CrossRef]
14. Bordoni, M.; Meisina, C.; Valentino, R.; Lu, N.; Bittelli, M.; Chersich, S. Hydrological factors affecting rainfall-induced shallow landslides: From the field monitoring to a simplified slope stability analysis. *Eng. Geol.* **2015**, *193*, 19–37. [CrossRef]

15. Ishaka, M.F.; Alib, N.; Kassimb, A.; Bahru, U.J. Analysis of suction distribution response to rainfall event and tree canopy. *J. Teknol.* **2016**, *78*, 83–87.

16. Heyerdahl, H. Influence of extreme long-term rainfall and unsaturated soil properties on triggering of a landslide—A case study. *Nat. Hazards Earth Syst. Sci.* **2017**. [CrossRef]

17. Mahmood, K.; Kim, J.M.; Ashraf, M. The effect of soil type on matric suction and stability of unsaturated slope under uniform rainfall. *Ksce J. Civ. Eng.* **2016**, *20*, 1294–1299. [CrossRef]

18. Yang, Y.; Wang, Y.; Wu, Y. The Influence of Rainfall on Soil Slope Stability. *Electron. J. Geotech. Eng.* **2015**, *20*, 13071–13080.

19. Wu, L.; Zhou, Y.; Sun, P.; Shi, J.; Liu, G.; Bai, L. Laboratory characterization of rainfall-induced loess slope failure. *Catena* **2017**, *150*, 1–8. [CrossRef]

20. Manenti, S.; Amicarelli, A.; Todeschini, S. WCSPH with limiting viscosity for modelling landslide hazard at the slopes of artificial reservoir. *Water* **2018**, *10*, 515. [CrossRef]

21. Khalid, M.; Kim, J.M. Effect of hydraulic conductivity on suction profile and stability of cut-slope during low intensity rainfall. *J. Korean Geotech. Soc.* **2012**, *28*, 63–70. [CrossRef]

22. Qi, S.; Vanapalli, S.K. Hydro-mechanical coupling effect on surficial layer stability of unsaturated expansive soil slopes. *Comput. Geotech.* **2015**, *70*, 68–82. [CrossRef]

23. Wu, L.; Xu, Q.; Zhu, J. Incorporating hydro-mechanical coupling in an analysis of the effects of rainfall patterns on unsaturated soil slope stability. *Arab. J. Geosci.* **2017**, *10*, 386. [CrossRef]

24. Zhang, G.R.; Qian, Y.J.; Wang, Z.C.; Zhao, B. Analysis of rainfall infiltration law in unsaturated soil slope. *Sci. World J.* **2014**, *2014*, 567250. [CrossRef]

25. Rahimi, A.; Rahardjo, H.; Leong, E.C. Effect of hydraulic properties of soil on rainfall-induced slope failure. *Eng. Geol.* **2010**, *114*, 135–143. [CrossRef]

26. Sasekaran, M. Impact of Permeability and Surface Cracks on Soil Slopes. Master's Thesis, The University of Manchester, Manchester, UK, 2011.

27. Gottardi, G.; Gragnano, C.G. On the role of partially saturated soil strength in the stability analysis of a river embankment under steady-state and transient seepage conditions. Presented at the E3S Web of Conferences. E3S Web of Conferences 9, Paris, France, 12–14 September 2016; Volume 19002. [CrossRef]

28. Zhai, Q.; Rahardjo, H.; Satyanaga, A. Variability in unsaturated hydraulic properties of residual soil in Singapore. *Eng. Geol.* **2016**, *209*, 21–29. [CrossRef]

29. Wu, L.Z.; Zhang, L.M.; Zhou, Y.; Li, B. Analysis of multi-phase coupled seepage and stability in anisotropic slopes under rainfall condition. *Environ. Earth Sci.* **2017**, *76*, 469. [CrossRef]

30. Ojha, R.; Corradini, C.; Morbidelli, R.; Govindaraju, R.S. Effective saturated hydraulic conductivity for representing field-scale infiltration and surface soil moisture in heterogeneous unsaturated soils subjected to rainfall events. *Water* **2017**, *9*, 134. [CrossRef]

31. Mahmood, K.; Ryu, J.H.; Kim, J.M. Effect of anisotropic conductivity on suction and reliability index of unsaturated slope exposed to uniform antecedent rainfall. *Landslides* **2013**, *10*, 15–22. [CrossRef]

32. Yeh, H.F.; Wang, J.; Shen, K.L.; Lee, C.H. Rainfall characteristics for anisotropic conductivity of unsaturated soil slopes. *Environ. Earth Sci.* **2015**, *73*, 8669–8681. [CrossRef]

33. Yeh, H.F.; Tsai, Y.J. Analyzing the Effect of Soil Hydraulic Conductivity Anisotropy on Slope Stability Using a Coupled Hydromechanical Framework. *Water* **2018**, *10*, 905. [CrossRef]

34. Chen, P.; Mirus, B.; Lu, N.; Godt, J.W. Effect of hydraulic hysteresis on stability of infinite slopes under steady infiltration. *J. Geotech. Geoenvironmental Eng.* **2017**, *143*, 04017041. [CrossRef]

35. Kim, J.; Hwang, W.; Kim, Y. Effects of hysteresis on hydro-mechanical behavior of unsaturated soil. *Eng. Geol.* **2018**, *245*, 1–9. [CrossRef]

36. Zhang, J.; Li, J.; Lin, H. Models and influencing factors of the delay phenomenon for rainfall on slope stability. *Eur. J. Environ. Civ. Eng.* **2018**, *22*, 122–136. [CrossRef]

37. Samadi, A.; Amiri-Tokaldany, E.; Darby, S.E. Identifying the effects of parameter uncertainty on the reliability of riverbank stability modeling. *Geomorphology* **2009**, *106*, 219–230. [CrossRef]

38. Samadi, A.; Amiri-Tokaldany, E.; Davoudi, M.H.; Darby, S.E. Experimental and numerical investigation of the stability of overhanging riverbanks. *Geomorphology* **2013**, *184*, 1–19. [CrossRef]

39. Fox, G.A.; Heeren, D.M.; Wilson, G.V.; Langendoen, E.J.; Fox, A.K.; Chu-Agor, M.L. Numerically predicting seepage gradient forces and erosion: Sensitivity to soil hydraulic properties. *J. Hydrol.* **2010**, *389*, 354–362. [CrossRef]

40. Zhong-Min, Y.; Yan-Qiong, G. Coupling effect of seepage flow and river flow on bank failure. *J. Hydrodyn.* **2011**, *23*, 834–840.

41. Yong, G.L.; Robert, E.; Thomas, Y.O.; Simon, A.; Blair, P.; Greimann, K.W. Modeling of multilayer cohesive bank erosion with a coupled bank stability and mobile-bed mode. *Geomorphology* **2014**, *243*, 116–129.

42. Duong, T.T. Assessment of Riverbank Stability—The Perspectives of Unsaturated Soils and Erosion Function. Doctoral Dissertation, Ibaraki University, Mito, Japan, September 2014.

43. Liang, C.; Jaksa, M.B.; Ostendorf, B.; Kuo, Y.L. Influence of river level fluctuations and climate on riverbank stability. *Comput. Geotech.* **2015**, *63*, 83–98. [CrossRef]

44. Vanapalli, S.; Fredlund, D.; Pufahl, D.; Clifton, A. Model for the prediction of shear strength with respect to soil suction. *Can. Geotech. J.* **1996**, *33*, 379–392. [CrossRef]

45. Van Genuchten, M.T. A closed-form equation for predicting the hydraulic conductivity of unsaturated soils 1. *Soil Sci. Soc. Am. J.* **1980**, *44*, 892–898. [CrossRef]

46. GEO-SLOPE International Ltd. Seepage Modeling with SEEP/W. Available online: http://www.geo-slope.com (accessed on 2 January 2019).

47. GEO-SLOPE International Ltd. Stability Modeling with SLOPE/W. Available online: http://www.geo-slope.com (accessed on 2 January 2019).

48. Krahn, J. The 2001 RM Hardy Lecture: The limits of limit equilibrium analyses. *Can. Geotech. J.* **2003**, *40*, 643–660. [CrossRef]

49. Aryal, K.P. Slope Stability Evaluations by Limit Equilibrium and Finite Element Methods. Doctoral Thesis, Norwegian University of Science and Technology, Trondheim, Norway, 2006.

MDPI
St. Alban-Anlage 66
4052 Basel
Switzerland
Tel. +41 61 683 77 34
Fax +41 61 302 89 18
www.mdpi.com

Water Editorial Office
E-mail: water@mdpi.com
www.mdpi.com/journal/water